Edouard Hugon

Cosmology

editiones scholasticae

Volume 30

Edouard Hugon

Cosmology

Translated, with Notes
by Francisco J. Romero Carrasquillo

editiones scholasticae

Bibliographic information published by Deutsche Nationalbibliothek
The Deutsche Nationalbibliothek lists this publication in the Deutsche Nationalbibliographie; detailed bibliographic data is available in the Internet at http://dnb.ddb.de

P.O. Box 15 41, D-63133 Heusenstamm
www.editiones-scholasticae.de

ISBN 978-3-86838-531-1

2013

Printed on acid-free paper

Printed in Germany
CPI buchbücher.de GmbH

INTRODUCTION TO SCHOLASTIC THOMISM
IN PREPARATION FOR THE THEOLOGY OF THE ANGELIC DOCTOR

II^A PARS:
NATURAL PHILOSOPHY

I^A-II^AE:

COSMOLOGY

On the World, with regard to its Efficient Cause;
On the World, with regard to its Material and Formal Causes;
On the World, with regard to its Final Cause.

By Rev. Fr. Edouard Hugon, O.P.
Professor of Sacred Theology, Angelicum College, Rome
Member of the Pontifical Academy of St. Thomas Aquinas, Rome

Translated, with Notes by:
Francisco J. Romero Carrasquillo
Professor of Philosophy, Universidad Panamericana (México)
Founder of the *Ite ad Thomam* Translation Proejct

Approbations

By order of our superiors we have read the work published by the Rev. Fr. Edouard HUGON entitled *Cursus Philosophiae Thomisticae*, for the use of students of Thomistic philosophy, which consists in six volumes. Since in it we found sound doctrine arranged in a convenient method, apt for the advancement of those who undertake the learning of the rational disciplines, and a true preparation for the acquiring of the more divine wisdom of the *Summa theologiae*, it pleased us to approve, *servandis de jure servandis*, the publication of this new edition.

Given at Rome, from the Angelicum College on June 24, 1920.

Fr. Reginald GARRIGOU-LAGRANGE, O.P.
S. Theol. Magister

Fr. Ceslaus POBAN-SEGOND, O.P.
S. Theol. Magister

Imprimatur:
Br. Buenventura GARCÍA DE PAREDES, O.P.
Magister Generalis Ord. Praed.

Imprimatur:
Paris, June 12, 1927.
V. DUPIN
Vicar. general.

Letter of His Holiness Pius X

Cursus philosophiae thomisticae

To my beloved son Edouard HUGON, *priest of the Order of Preachers, Doctor in Theology at the Pontifical Angelicum College.*

Pius PP. X

Beloved Son,

Greetings and Apostolic Blessings.

If the teaching of Aquinas ought anywhere to be held in the highest honor and zeal, it certainly ought to be so held by that religious family of which he was the particular light and ornament. And indeed among the members of the distinguished Dominican Order are esteemed in our memory many famous men who follow in the footsteps of such a great Doctor—a Doctor who never grows old—and who undertake to shed a brilliant light on Christian dogmas and teachings, and irrefutably to defend them.

Your *Cursus philosophiae thomisticae*, which We recently received from you as a gift offered in filial affection, demands that you, beloved son, be held not least among this number. For the judgments of men who are well versed in philosophy concerning you are well known, men who praise in your volumes the unadulterated teaching of Saint Thomas, and the wealth and coherence of your arguments and the clarity of your style, and who praise you particularly for applying ancient scholastic principles to shed light on new advances in philosophy and to the judicious refutation of errors.

For this reason We thank you for the fruits of your labors which you have presented to Us as a mark of respect, but much more for the salutary assistance you have given to young men engaged in sacred studies. For, as We have often already said, nothing is so important for the advantage of the Church as that the wisdom of the Angelic Doctor should preside over the weighty studies of the young clergy, and we rejoice that this occurs so fruitfully in the case of students of sacred studies who have you either as a teacher or as an author.

As assurance of divine reward and a witness to Our paternal benevolence We most lovingly bestow our Apostolic Blessing upon you, dear son.

Given in Rome at St Peter's on the 16th of July 1913 in the tenth year of Our Pontificate.

Pius PP X

THE AUTHOR DEDICATES
THIS TREATISE ON THE WORLD
TO THE KING OF THE AGES,
THE IMMORTAL, INVISIBLE,
AND ONLY GOD:
THE FATHER, FROM WHOM ALL THINGS ARE,
THE SON, THROUGH WHOM ALL THINGS ARE,
AND THE HOLY GHOST, IN WHOM ALL THINGS ARE.

TRANSLATOR'S ACKNOWLEDGMENTS

This volume is the first fruit of the *Ite ad Thomam* Translation Project, founded in 2009. My heartfelt thanks to the project's collaborators: Daniel Torzala, Alan Aversa, Fr. Edward van den Bergh, Anselm Keiser, Lucas Alcole, Eli Rotello, Zachary Thomas, and many other friends who graciously volunteered to proofread different sections of this translation and to help with some difficult issues in translation. Also, special thanks to Alexandre Pinheiro, who supported us during the Project's first years, making it financially possible for us to work on the translation, as well as to my wife Hilaire who inspired and encouraged me throughout the endeavor with much patience and love.

TABLE OF CONTENTS

SECOND ARTICLE

Whether the world is a contingent being

THIRD ARTICLE

On pantheism

SECOND QUESTION

On the true origin of the world

FIRST ARTICLE

Who is the author of the world

SECOND ARTICLE

Whether the world is from God by way of emanation

THIRD ARTICLE

Whether the world is from God through creation

FOURTH ARTICLE

On the subject and the object of creation

FIFTH ARTICLE

Whether creation is a free act of God

THIRD QUESTION

On the Duration of the World

FIRST ARTICLE

Whether the world could have existed from eternity

SECOND ARTICLE

How Old is the Universe; or on Cosmogony and Geogony

THIRD ARTICLE

Whether the world is to last forever

SECOND TREATISE

ON THE WORLD WITH RESPECT TO ITS MATERIAL AND FORMAL CAUSES

FIRST QUESTION

On the constitutive principles of bodies in general

FIRST ARTICLE

The conditions of the first principles of bodies

SECOND ARTICLE

Atomism explained and refuted

THIRD ARTICLE

Dynamism explained and refuted

FOURTH ARTICLE

The existence of matter and form is proven

SECOND QUESTION

On the constitutive principles of bodies in particular

FIRST ARTICLE

On prime matter

SECOND ARTICLE

On substantial form

THIRD ARTICLE

On natural composites

FOURTH ARTICLE

Explanation of many things that are necessary for the complete understanding of hylemorphism

FIFTH ARTICLE

The history of hylemorphism is briefly reviewed

THIRD QUESTION

On the properties of bodies

FIRST ARTICLE

On quantity itself

SECOND ARTICLE

On the continuum

THIRD ARTICLE

On place

FOURTH ARTICLE

On space and the void

FIFTH ARTICLE

On duration in general

SIXTH ARTICLE

On time

SEVENTH ARTICLE

On the qualities of bodies

FOURTH QUESTION

On the principle of individuation

FIRST ARTICLE

On the individuation of bodies

SECOND ARTICLE

Application of the aforesaid doctrine to the individuation of the human soul and to the individuation of accidents

THIRD TREATISE

ON THE WORLD INSOFAR AS IT IS ORDERED TO AN END

FIRST QUESTION

What Nature Is

FIRST ARTICLE

On the notion of nature

SECOND ARTICLE

On motion

THIRD ARTICLE

On art and violence

SECOND QUESTION

How Nature is Ruled or Governed

FIRST ARTICLE

On the laws of nature

SECOND ARTICLE

On the nature of miracles

THIRD ARTICLE

On the possibility and knowability of miracles

THIRD QUESTION

What nature is for

FIRST ARTICLE

The end of nature

SECOND ARTICLE

Whether evolution is in agreement or in disagreement with the end of nature

End of the First Part of Natural Philosophy.

NATURAL PHILOSOPHY

FIRST PART

PROLEGOMENA

PREPARATION FOR NATURAL PHILOSOPHY

I. – THE TERM "WORLD." In order to understand more fully the notion of Cosmology, or of the science of the world, it is necessary first to explain what the term "world" or *cosmos* means. *Cosmos* is a word first used by Pythagoras, meaning ornament, or elegance, which the Latins translated as *mundus* [and which is in turn translated into English as "world"]. "That which the Greeks called *cosmos*, the term for ornament, we call *mundus* in reference to perfect and absolute elegance."[1] As far as its etymology, therefore, *mundus* means elegance and beauty. For this reason, God, who is the first and highest beauty, can especially be called *mundus*. For He is the archetypal world, in whom preexist all the ideas of things, and from whom all created things derive their elegance and beauty. Of Him Boethius says, "You Yourself being most fair, You bear a fair world in Your Mind and form it in the same likeness..." (*Pulchrum pulcherrimus ipse mundum mente gerens similique ab imagine formans*).

II. – THREE SENSES OF 'WORLD'. But outside of God, three worlds can be distinguished, which deserve to be called 'elegant' and

[1] PLINIUS, *Hist. natur.*, 2.4: "Quem Greci *kosmon* nomine ornamenti appellavere, eum et nos a perfecta absolutaque elegantia mundum dicimus."

'beautiful', namely, the bodily world, the human world, and the angelic world. Now, even greater than beauty is splendor. And in this triple world a manifold splendor shines out: in the corporeal world the splendor of natural form and a certain vestige of the Blessed Trinity shine out, as we shall explain in the treatise On the Beautiful; in the human and angelic worlds the splendor of natural form, the splendor of grace or glory, and even the divine splendor, which comes from the indwelling of the Blessed Trinity, shine forth. Also, other elements of beauty are present. For a wondrous variety is found in the corporeal world: that is, a variety of beauty, such as light, flowers, stars and the other beautiful things in the sky, the earth, and the sea; a variety in sublimity, as the vastness of the sea, the immensity of the skies, etc. In the human world variety is found, both in the parts of the body and in the powers of the soul. Now, the angelic world shows forth an almost infinite variety, for in angels there are as many species as individuals. Finally, in each of these three worlds concurs a unity, both with regards to its order and to its end; and among themselves these three worlds are connected by a twofold link, namely, a dynamic link, or link of causality, and a teleological link, that is, a link of finality. This threefold sense of 'world' constitutes the entirety of created things and is often referred to by the name of "Universe."

III. – CHRIST, THE PERFECT WORLD.[2] For the sake of beauty, it is fitting that there be a certain world that is a sort of compendium or recapitulation of the others. And man is in some way the compendium of all the worlds. "For he has *esse*[3] in common with stones, living in

[2] The points we make here are, of course, extraneous to Natural Philosophy, but they are nonetheless useful for one to know, in order that we may understand more fully the general notion of "world" which we are explaining. Thus it will appear how fitting the Incarnation is and how profound is the harmony between Philosophy and Theology.

[3] *Esse*, that is, the being or existence of a thing. Understood metaphysically, *esse* is the act which, together with essence, make a concrete existent thing. *Esse* is thus contrasted with *ens*, '*a* being', the thing itself that has *esse*. Both Latin terms, *esse* and *ens*, can be translated into English as 'being'; as a result, the English term 'being' can be ambiguous. In order to avoid such ambiguity, we have left the Latin

common with trees, sensing in common with animals, and understanding in common with angels. If man, therefore, has something in common with all creatures, then in a sense man is every creature."[4] For this reason man is called a *microcosm*, a small world; imperfectly so, however, for the soul holds the lowest place among spiritual things... There would need to be some world that is both a body and God, man and God, spirit and God. Now, that world is not only an ideal, but a reality: Jesus Christ. In Him, as the Apostle testifies, God renewed all things, or as the Greek text has it, God made a *recapitulation of all things* (*anakephalaiôsasthai*).[5]

He first recapitulated the material world. The human body is the ideal among inferior bodies; but the exemplar of the human body is the body of Jesus Christ. In the body of his Son, therefore, God renewed all things. Now, in the soul of Christ are recapitulated both the human world and the angelic world: for his most holy Soul has all the perfections of all men and gathers within itself greater knowledge and grace than all the angels together. In Christ, therefore, the three worlds are summarized; they are, in fact, joined together with the divine and archetypal world itself, and form one world with it through the hypostatic union. For, in Christ, Divinity, soul, and body subsist in one Person. Christ, therefore, is the most perfect world, in which all the worlds are made one—one, that is, in the Person.

IV. – The World of Which we Shall Here Speak. Cosmology does not inquire concerning the archetypal world [i.e., God], which is the object of Theology.

In the absolute sense, Cosmology is the science of the entire world, insofar it includes all creatures, bodies, man, and spirit. It has become customary, however, for the term Cosmology to be used to

---

terms *esse* and *ens* untranslated for the most part; where we have translated the, we have included the original Latin in parentheses. – ***The Translator.***

[4] St. Gregory the Great, *Homil. de Ascensione* (*PL* 76, 1274): "Habet namque commune esse cum lapidibus, vivere cum arboribus, sentire cum animalibus, intelligere cum angelis. Si ergo commune habet aliquid cum omni creatura homo, juxta aliquid omnis creatura est homo."

[5] Cf. Ephesians 1:10: "[I]nstaurare omnia in Christo...."

denote, in the proper sense, the science of this sensible, corporeal, and mobile world, even though it does examine those things that are common to the triple world, namely, the angelic, human and corporeal, such as those things that regard creation. This understanding of Cosmology pertains to Physics, or the Natural Philosophy of the ancients.

V. – THE OBJECT OF COSMOLOGY, OR NATURAL PHILOSOPHY.[6] There is more than one manner of speaking among Scholastics. For, as John of St. Thomas testifies, "certain authors assert that the object [of Natural Philosophy] is mobile being (*ens mobile*),[7] others mobile body (*corpus mobile*), others natural or sensible body (*corpus naturale, vel sensibile*), although others say that all of these coincide in reality and differ only in the manner of speaking."[8] Our position is that of St. Thomas: "The subject of [natural] philosophy is mobile being *simpliciter*."[9] We shall settle the issue briefly.

(1) The material object (*objectum materiale*)[10] is natural body (*corpus naturale*), or the sensible world. For all agree that this part of

[6] For the sake of clarity, I took the liberty of slightly re-arranging a few paragraphs in this dense, yet important section; I left their content intact. – ***The Translator.***

[7] Hugon is utilizing the term "mobile" in its technical, Aristotelian and Thomistic sense, as meaning "changeable." Similarly, "motion" here means any type of accidental change—not just locomotion, or the change of place. – ***The Translator.***

[8] *Phil. Nat.* q. 1, a. 1: "Quidam asserunt objectum esse ens mobile, quidam corpus mobile, quidam corpus naturale, vel sensibile, licet haec omnia dicant alii in idem coincidere et solum differre penes modum loquendi."

[9] ST. THOMAS, *In Phys.*, proemium: "Subjectum philosophiae est ens mobile simpliciter." *Simpliciter*, "simply [speaking]," as opposed to *secundum quid*, "according to something" ("in a certain sense"). – ***The Translator.***

[10] The material object of a science, faculty, or power is that which the science, faculty, or power considers. For example, the material object of anatomy, physiology, and medicine is the human body; similarly, both the faculty of sight and the intellect can have the same tree as their material object, insofar as the sight sees the tree and the intellect thinks of the tree. The material object is distinguished from the *formal* object, in that the concept of formal object adds the *aspect* of the thing that is known. Thus, for example, the formal object of anatomy is the *structure* of the human body, that of physiology is the *function* of the human body,

philosophy does not deal with separate substances, but with these mobile beings which do not possess the *ratio*[11] of incorruptible beings, and which constitute the material world.

(2) Now, **a formal object** is that which a faculty considers directly and *per se* in the material object. But that aspect of the world which Natural Philosophy examines is motion, whether motion in general, or the motion of generation and corruption, or vital motion, and for this reason it is distinguished from Metaphysics. **(a) The 'formal object which' (*objectum formale quod*)**[12] is 'mobile being insofar as it is mobile' (*ens mobile ut mobile*). Note that 'natural being' and 'mobile being' coincide in the reality. Natural being, as the word indicates, is something that consists in a nature; but here 'nature' is taken insofar as it is the principle of motion and rest. Therefore, 'natural being' (*ens naturale*) designates that which has in itself a principle of rest and motion. 'Mobile' means the same here; for it is not taken as something proximately or formally mobile, but only mobile by aptitude, or in other words, that which has in itself a principle of motion. It is better to say

---

and that of medicine is the *healing* of the human body. Similarly, the material object of the faculty of sight may be that tree, but its formal object is that tree's *color*; the material object of the intellect may also be that tree, but its formal object is that tree's *essence*. – ***The Translator.***

[11] Throughout the work I have left the word *ratio* untranslated. It is a very versatile term that can be variously translated, depending on context, as "reason," "nature," "concept," "aspect," "argument," etc. The term is ubiquitous in Hugon's manual, as well as in the entire Thomistic tradition; the beginning student should aim to get a sense of its latitude of meaning from context. In this context, it means "nature": mobile beings do not possess an incorruptible nature/*ratio*. – ***The Translator.***

[12] The *objectum formale quod* of a science, faculty, or power is the complete formal object, which includes both the material object and the aspect formal object. Thus, the *objectum formale quod* of sight is a physical object *qua* colored, and that of the intellect is the physical object *qua* intelligible; similarly, the *objectum formale quod* of Natural Theology is God *qua ens primum*, or as the first cause of being, whereas the *objectum formale quod* of Sacred Theology is God *qua* God. The *objectum formale quod* is distinguished from the *objectum formale quo*, which, as Hugon explains below, is the "medium," the principle or source, of knowledge. For example, in Sacred Theology the *objectum formale quo* is Divine Revelation, whereas while doing Natural Theology, the *objectum formale quo* is human reason. – ***The Translator.***

'*mobile* being' (*ens mobile*) rather than '*natural* being' (*ens naturale*); for although the sense is the same, 'mobile' expresses the object better than 'natural'. It is also more correct to say 'mobile *being*' rather than 'mobile *body*' (*corpus mobile*), because it is not yet known whether that which moves is a body; but *a priori* it is plain that every mobile thing is an *ens*. The "mobile being" of which we treat here is not only that which is substantially mobile [i.e., which can undergo substantial change], but also that which is accidentally mutable [i.e., which can undergo accidental change]; substance, however, is *subjectively* mobile, for the first *subject* of motion is a nature, which necessarily subsists *per se*. **(b) The 'formal object under which' (*objectum formale sub quo*)** is the medium (*medium*), or the principle, by which Natural Philosophy illustrates its conclusions and makes demonstrations, which is the degree of abstraction from singular matter. In order to understand this, recall what was said in *Major Logic*.[13]

VI. – Natural Philosophy is a science[14]; that is to say, it does not only assert its conclusions, but rather proves them demonstratively from causes. The fact that it is most useful and noble shines forth from similar evidence, since nature is like a book in which we read the divine ideas. For the natural order of knowledge is such that we rise through the visible world, which is better known to us, to the world of our soul and to the archetypal world. It is distinguished from the other physical sciences, even though it coincides with them in its material object. For these touch on the proximate causes, or certain phenomena of bodies, but Natural Philosophy treats of the world through the highest causes, and seeks and finds the Alpha and Omega, the beginning and end of the universe.

---

[13] Cf. Treatise III, q. 1, a. 1.

[14] Here, Hugon has in mind the Aristotelian-Thomistic notion of science as "knowledge through causes" (*cognitio per causas*). A "cause" in this sense, is the '*be*-cause' of a thing, the thing's *raison d'être*, its reason for being. Thus, a science, in the Aristotelian sense, gives reasons for why a thing is, and why it is the way it is. In this sense, natural philosophy deserves to be considered a science, because it gives reasons for why physical (or mobile) things are. – ***The Translator.***

VII. – Division of Natural Philosophy. The ancients, in view of the formal object, which is *mobile* being, divided this science into four parts: (1) On mobile being in general; (2) on mobile being as far as its locomotion; (3) on the motion of generation and corruption; (4) on vital motion. But under the above division they introduced many topics that today are taught in the physical sciences, outside of Philosophy.

Many moderns take their division from the four genera of causes, and treat of Natural Philosophy thus: (1) On the world with regard to its efficient cause, or on the characters, on the origin of the world, etc.; (2) on the world with regard to its material and formal causes, or on the constitution of bodies; (3) on the world with regard to its final cause, and therefore on the nature or end of the world and its laws.

No one will deny that this partition is the best. And let us recall that within Natural Philosophy not only Cosmology is included, but also the first part of Psychology. Indeed, Natural Philosophy examines things as abstracted from singular matter, but not as abstracted from sensible matter. But the vegetative soul, the sensitive soul, and even the human soul insofar as it is the form of the body, or insofar as it gives man his vegetative and sensitive being, are not abstracted from all sensible matter. For this reason the treatise on the soul pertains to Natural Philosophy.

Therefore, we divide Natural Philosophy into two volumes:

The first volume (*Ia-IIae*) considers inanimate natural being (*de ente naturali inanimato*), although many things shall be considered which are true of all being: and this is Cosmology, strictly speaking. This first volume contains three treatises:

(1) On the world with regard to its efficient cause;

(2) On the world with regard to its material and formal causes;

(3) On the world with regard to its order and its laws, or on nature.

The second volume (*IIa-IIae*) considers animate being (*de ente animato*), or the soul: the first part of Psychology. It contains three treatises as well:

(1) On the soul, or life in general;

(2) On vegetative and sensitive life;

(3) On the human soul as united to the body, considered both as far as its substance and as far as its faculties and operations.

VIII. – The fortune of Natural Philosophy. The philosophy of nature has existed from the beginning, at least in a certain way, for it is proper to man that he enquire and probe the visible world, discerning its causes. Among the ancients the naturalistic school is well known, which attended especially to the investigation of natural principles, although it indulged too much in materialism. Aristotle explained the entire system of Natural Philosophy in his *Physics*, *On the Heavens and the World*, *On the Soul*, and in the diverse books *On Animals*. The others schools were not strangers to this study, although many concentrated especially on the study of man—their principle was: "Know thyself." The Fathers of the Church often argue against the pagans from the viewpoint of Nature. But the Scholastics consider Natural Philosophy as a special science, whose task is to treat of mobile being in general, on mobile being as far as its locomotion, its motion of generation and corruption, and finally its vital motion. In modern times, however, the natural sciences having been wondrously developed, Philosophy was banished from Physics; hence it happens that scientists, in exploring facts or at least aiming at proximate causes, have neglected ultimate causes and first principles. For this reason, a philosophical regard in the study of nature is more than ever desired today. This science, however, was not wholly neglected by the Moderns. But the other part of Natural Philosophy, which is *Psychology*, is especially cultivated by modern authors, as if it were a primary and principal science.

IX. – At what Point in the Course of Philosophical Studies should Natural Philosophy be Treated. The place that is to be accorded to this science is immediately after Logic. For the order of treatment should correspond to the degree of abstraction, so that knowledge be led from the better known to the unknown.

But the first degree of abstraction is [the one used] in Natural Philosophy, which examines the visible world, while Ontology is about more abstract things, which are at the apex of immateriality. Therefore Natural Philosophy is prior to Ontology. Now, because animate being is more abstract than inanimate being, that part which examines inanimate being, or simple Cosmology, must precede that part which treats of animate being, namely, Psychology.

Many recent authors, following Cousin, begin with Psychology, which, according to them, is to be treated at the threshold of Philosophy, before Logic itself. But they are mistaken. For, since the soul is not known intuitively or through its own concept, but the concept depends on abstract things, its study requires many previous things that are dealt with in the Philosophy of Nature and in Logic. Thus we recognize the soul through apprehension, judgment, and reason, the notions of which depend on Logic. Further, Logic supplies the manner of proceeding and of making demonstrations in Psychology; but the laws of Logic, for example, about propositions and about reasoning, do not depend on the nature of the soul being known beforehand. Logic, therefore, is placed before Psychology, and not Psychology before Logic. We, however, assign Psychology, as far as its first part, to Natural Philosophy; but as far as its second part, which examines the intellect and the will, to Metaphysics. Thus, we place [Psychology] between Cosmology and Ontology.

X. – AUTHORS WHO HAVE WRITTEN ABOUT NATURAL PHILOSOPHY. The first to be consulted is ARISTOTLE: the eight books of the *Physics*, the four books *On the Heavens*, the five books *On Generation and Corruption*, the four books of *Meteorology*, and the three books *On the Soul*. See also Pliny, *Natural History*.

Also to be consulted: St. ALBERT THE GREAT, who wrote wonderful works on natural things; VINCENT OF BEAUVAIS, *Speculum naturae*; St. THOMAS, *Commentaries on Aristotle*, *Summa contra gentiles*, *Summa theologiae* Ia; ROGER BACON, who successfully anticipated the findings of many recent authors; TOLEDO, the CONIMBRICENSES, and the COMPLUTENSES, *Commentaria in libros Aristotelis*; JOHN OF ST. THOMAS, ALAMANNUS, GOUDIN, in their *Physics*; GUERINOIS, *Clypeus philosophiae thomisticae*, Physics.

Francis Bacon, in his *Instauratione magna*, applied an immoderate sort of empiricism to the physical sciences; his works contain numerous errors. J. De Maistre, *Examination of the Philosophy of Bacon*, and Justus of Liebig have written against him. In modern times, many works on Natural Philosophy have come out, of which many will be cited in the course of our study.

XI. – The Method to be followed in Natural Philosophy. Since this discipline intends to provide certain knowledge by examining the visible world, it is necessary for us to rely upon experience, together with universal principles, analysis, and synthesis. It must give itself to induction, the physical sciences, and the most recent findings of the scientists, so that, by a gradual process, the mind may move from sensible things to intelligible things; but it must apply abstract principles, so that it may obtain necessity and the certitude that is free from deception. Recall what we discussed concerning the method in *Major Logic*.

FIRST TREATISE

ON THE WORLD
WITH RESPECT TO ITS EFFICIENT CAUSE

When something proceeds from its efficient cause, it arises from something from which it derives its existence. Therefore, to inquire whether the world has an efficient cause is to ask whether it be derived from another [thing] (*ab alio*). In order that it may be clear whether the world has in itself or in another the *ratio* of its existence, we must investigate which are its essential characters, namely, whether it is simple, or composed, whether it is necessary, or contingent, distinct from God, etc.

FIRST QUESTION

On the Peculiar Characters of the World

FIRST ARTICLE

WHETHER THE WORLD IS SIMPLE AND SOMETHING PER SE ONE OR SOMETHING COMPOSED

I. – Monism. Here we attack the doctrine of the Monists. *Monism* is the opinion that asserts that all the things in which the world consists are in reality a single being, which nonetheless appears diverse due to its diverse modifications and evolutions.

Monism is twofold: *absolute* and *moderate*. Absolute monism claims that everything is a simple being, and that there are no individuals that are really distinct from others.

But the moderate monist view is that the world, although perhaps not an altogether simple being, is nonetheless one, not merely with an accidental unity, but with a substantial unity, i.e., it is *per se* one (or *unum per se*).

II. – Defenders of Monism. Among the ancients, the Eleatics and Stoics especially adhered to monism. The Eleatics, such as Xenophon and Parmenides, thought that the world is one eternal being, existing as a round sphere, and that distinction and multitude are only an illusion of the senses.

The Stoics, such as Zeno and Chryssipus, posited two principles: one passive and material (*to paschon*); the other active and intelligent (*to poioun*), which are related to each other as body and soul, producing a single substance. Therefore, according to this opinion, God, or the active principle, is the soul of matter; and just as, from the soul and the body, a thing that is *per se* one is composed, so from the soul of the world and its matter results something that is *per se* one; hence they speak of a being that is both one and all things: *one and all* (*hen kai pân*).

The Hindus advanced a certain monism that is [reducible to] emanationism; we shall speak of it below. The Neo-Platonists also adhere to an emanationist monism.

Among the moderns, the following profess the doctrine of monism: (1) The pantheists, who contend that all things are God, and God is all things. This was taught by Baruch Spinoza in the Seventeenth Century; more recently Cousin gives occasion to pantheism when he calls the world a divine and necessary substance. Vacherot presents the world and God as one object that is considered by the mind under a twofold aspect.[15] (2) The German Transcendental Philosophers, who claim that the thinking subject makes all things (which doctrine is often called *Solipsism*); or that all things are the ego itself, or a pure idea. (3)

---

[15] Vacherot, *Histoire critique de l'École d'Alexandrie*, III, p. 479: "There exists only one, simple, immutable, infinite, universal substance, of which the substances that are called 'individuals' are only determinations" (*Il n'existe qu'une substance simple, immuable, infinie, universelle, dont les substances dites individuelles, ne sont que des déterminations*).

Many scientists believe that the world possesses that unity which is proper to an organism, and that the world makes itself evolve through reason in the manner of an organism.

Many reject any distinction between living and non-living things. They teach that all things, even living things, proceed from a single primitive monera, which by evolving from an inorganic state arrived at an organic and animate state.

Huxley and Heckel held that they had found this primitive and universal monera in a certain sticky substance extracted from the sea, which they called *Bathybium*; but *Bathybium* brought about only deceit and ridicule.[16]

(4) According to Schopenhauer, the world is reducible to a single, blind volition. "On account of this great misfortune, and out of desperation concerning its unhappy fate and great fault, this [volition] evolves into the sensible world (for, the world having come about, how can one explain it any other way than through some infinite need?) and then, disgusted by its miserable existence, it desires the rest of nothingness (Nirvana)."[17]

III. – THE THREE FORMS OF MONISM. From what was said above it is clear that monism can take three forms: *Pantheism* (*pan-*, "all things"; *theos*, "god"), *Panthelism* (*pan-*, "all things"; *thelesis*, "will"), and *Panhylism* (*pan-*, "all things"; *hyle*, "matter"). *Pantheism* claims that all things are one; *panthelism* claims that all things are one will, which is blind and fatal; and *panhylism* claims that all things are a single, material thing. It is not necessary to consider each of the forms of this error; it is sufficient to refute all of them at once with general arguments.

---

[16] Cf. DUILHÉ DE SAINT-PROJET, *Apolog. Scientifique. Orig. et dévelop de la vie.*

[17] Cf. PESCH, *Philosophia Naturalis*, n. 274: "Haec ob infortunium ingens et prae desperatione de sorte sua infelicissima et gravi culpa sese evolvit ad mundum adspectabilem (nam ortus mundi, quomodo explicari posset quam ex indigentia aliqua eaque infinita?) et deinde pertaesa miserabilem suam existentiam, quietem appetit ipsius nihil (Nirwana)."

IV. – First Conclusion[18]: "Monism, in whichever form it be proposed, is self-contradictory."

Proof.[19] Monism, in whichever form it be proposed, results in this, that the entire world is one substance, or a single principle. But this of itself implies that all the things in which the world consists are a single substance. Therefore, in whichever form it be proposed, monism is self-contradictory.

Proof of the Minor. 1st Argument. – If all things are one substance and one principle, then body and spirit, living and non-living, man and non-man, virtue and vice, finite and infinite, would be one and the same thing. But all these things are related to each other as contraries or contradictories. Therefore, those who claim that all things are a single substance, by that very fact conclude that contraries and contradictories are the same, which is to discard the first principle of all things. ***Objection.*** The aforesaid determinations are diverse respects of the same substance. But it is not contradictory for one and the same thing to receive contraries or contradictories according to diverse respects. *Therefore.*[20] ***Reply.*** *I deny the major* [*premise of the objection*].[21] They are not something relative, but something absolute; because they are intrinsically identical with substance, such that it is true when one asserts, "this substance is intrinsically alive; this substance is intrinsically non-living; this substance is intrinsically a man; this substance is intrinsically a non-man." Therefore, the simultaneous affirmation and negation of the same determination, for example, the simultaneous affirmation and negation of living, or of man, falls within

---

[18] A "conclusion" means a thesis to be defended. – ***The Translator.***

[19] For the structure of a "proof" in Hugon's work, see the Introduction. – ***The Translator.***

[20] Hugon, following scholastic custom, often omits the conclusion of the argument and simply says "Ergo," assuming that the reader will know what is the point being argued. In this case, since the argument is an objection against the "proof of the major," the conclusion being omitted is the opposite (or the denial) of the conclusion of the proof of the major, that is, the denial of the claim that "those who claim that all things are a single substance, by that very fact conclude that contraries and contradictories are the same." – ***The Translator.***

[21] The minor premise, which contains the minor term of the argument or objection, is usually the second premise in the argument (and is, thus, preceded by the major premise and followed by the conclusion). – ***The Translator.***

the substance itself, considered absolutely and intrinsically. But the simultaneous affirmation and negation of the same thing that falls in the same substance, considered intrinsically and absolutely, implies a contradiction. *Therefore.*

2nd Argument. In whatever beings the first principle of operation is multiple and diverse, the natures of the beings will themselves be multiple and diverse, since the first principle of operation is nature, or substance. And in the world, the first principles of operation are multiple and diverse. Therefore, natures, or substances, are multiple and diverse. ***Proof of the Minor.***[22] The first principles of action are diverse and distinct if actions themselves are discrete and wholly independent of other actions. This is evidenced through experience. Thus, the action and activity of a material molecule is altogether distinct and independently of the operations of a plant, hence, the plant having been destroyed, the activity of the material molecule does not thereby cease. Further, the energy of the plant is separate from the activity of animals; the or destruction of plants depends very little on the evolution or destruction of animals. Animals receive many and diverse cognitions, and many and distinct affections or passions. Men, as agents, are altogether distinct from other agents and from one another, both in cognition and in volition. Therefore, agents as agents are many and distinct. Therefore, it is altogether necessary that beings as beings be many and distinct. ***The argument can also be formulated thus***: If all things are one substance, then any singular being, being a part of the whole, will be partial and incomplete, and on that account no action will be complete or have a terminus (or endpoint), but, being incomplete, will be referred and ordered to another. But it is most evidently certain that many actions are whole, and of themselves have a terminus, as my intellection and volition have a terminus in themselves, and do not await another; and also as the action of the plant, or the actions of the animal are whole, independent, have a terminus, and do not depend on my volition. *Therefore.*

---

[22] Note the chain in arguments: this is the proof of the minor premise of the 2nd Argument, which itself is an argument in favor of the minor premise of the main "proof." – ***The Translator.***

3rd Argument. – Monism, which claims that matter becomes all things and that life is reduced from matter, contradicts (1) the certain facts of physics, which establishes the Law of Inertia [i.e., Newton's First Law of Motion] as unshaken; (2) the certain facts of biology, which proves, according to Pasteur, that spontaneous generation does not occur. Monism, therefore, is both philosophically and scientifically inconsistent.

V. – OBJECTIONS.[23]

1st Objection. The parts that are ordered to a whole constitute one thing. But single beings are parts of the world, which are ordered to the whole. All things, therefore, form together one thing.

Reply. *I distinguish the major.*[24] That the parts which are ordered to some whole constitute one thing, either *per se* one or *per accidens* one, I concede. *I subdistinguish "per se one"*: if they are parts that tend to one essential whole, then I concede [the premise]; if they are parts which conspire to form one collective whole, then I deny [the premise] *and I deny the consequence.*

I contradistinguish the minor. – That the single beings of the world are parts that are ordered to one collective whole, I concede; but that they are ordered to some essential whole, I deny.

I distinguish the conclusion[25]: That, therefore, the world is one by an accidental unity, I concede; but that it is one by a substantial unity, I deny. Certainly, the parts of an essential whole, such as matter and form, the soul and the body, constitute something that is *per se* one; but the parts of a collective whole, such as the stones that make up a house form something that is at least *per accidens* one. Yet the single beings of the world are parts of a collective whole only.

---

[23] These are objections, not against the arguments he just presented, but directly against the conclusion (or thesis). – ***The Translator.***

[24] For an explanation of the process of distinguishing, subdistinguishing, and contradistinguishing a premise, using this reply as an example, see the Introduction. – ***The Translator.***

[25] A distinguished conclusion means that Hugon accepts it in one sense (usually not the sense in which it was meant), and rejects it in another (usually the sense in which it was meant). – ***The Translator.***

2nd ***Objection***. We experience not only a tendency to the individual good, but also to the universal good. And such a tendency to the universal good requires a universal soul from which it proceeds. There is in the world, therefore, a universal soul, which together with matter makes up something essentially one. ***Reply***. *I deny the minor*. It is not necessary to posit a universal soul; it is sufficient to say that God, the maker of universal nature, has endowed us with such a tendency.

It is established, then, that the world is not a simple being; now we prove positively that the world is in many ways composed.

VI. – Second conclusion: "All possible composition is to be ascribed to the world."

Proof. Every possible composition is reducible to five genera, namely, (1) *essential*, (2) *entitative*, (3) *integral*, (4) *accidental*, and (5) *numeric* composition. And this fivefold composition is ascribed to the world. Therefore, every possible composition exists in the world.

Explanation of the Major. The kind of composition that is made out of essential and physical parts, such as out of soul and body, or out of matter and form, is called (1) *essential* composition. (2) *Entitative* composition results from that which is (*eo quod est*) and that by which something is (*eo quo est*), or from essence and existence. (3) *Integral* composition is that whose parts—thought they do not constitute the essence of the thing—nonetheless constitute the quantity of the whole or are required for the natural integrity of the whole, as was said in Logic. (4) *Numeric* composition emerges from parts that are in themselves complete and that are ordered to a certain *collective* whole only, but not to form one thing *per se*, such as a house. (5) *Accidental* composition is the result either of the union of many accidents or of a substance and its accidents.

Proof of the Minor. – In the world, we find: (1) *Essential* composition. A being that does not have essential parts is not subject to corruption, since corruption is the separation of essential parts. But corruptions occur in the world, as is evident when a plant becomes ashes and an animal becomes a carcass. In the world, therefore, there are essential parts and, thus, essential composition. The matter will become clearer when we discuss matter and form. (2) *Entitative* composition. Whatever does not exist by its own power is composed of essence and *esse*. And worldly substances do not exist by their own power, as we shall soon show. Worldly substances, therefore, are composed of essence and *esse*. Therefore, there is entitative composition in the world, or in worldly substances. (3) *Integral* composition. This is an effect of essential composition. Moreover, it is clear that extension and quantity are found in the world, and that there are parts that are required for the integrity of a whole, like the arms and legs in animals, etc. (4) *Numeric* composition. For we see that each being in the world is distinct from the rest, and that there are parts that are ordered to forming a collective whole. (5) *Accidental* composition. There are physical accidents really distinct from the substance in which they inhere, and which exist in the substance, and which form with it something that is *per accidens* one, as we shall show in *Metaphysics: Ontology II*, Treatise III. The world, therefore, is composed on all counts. This is the common opinion.

SECOND ARTICLE

WHETHER THE WORLD IS A CONTINGENT THING

I. – MEANING OF THE QUESTION. Contingent being means that which can be or not be without any contradiction, which therefore does not have in itself the *ratio* of its existence, but depends on another (*ab alio*). The question, therefore, "Whether the world is a contingent or necessary being" means "Whether the world exists through another (*ab alio*) or through itself (*a seipso*)."

II. – ERRORS. The false opinions that hold that the world exists through itself are reducible to the following systems: the system of an *infinite series of causes* and of *eternal matter*, the system of *fortuitous interaction of atoms*, the system of *active evolution, hylozoism*, and of *logical evolution*; and finally *pantheism*.

The first opinion teaches that there is an infinite series of beings, that each being has the *ratio* of its existence in another prior being, and so on *ad infinitum*; not that there is a first cause outside of the series, but that, even though each being is contingent, the whole collection of all things taken together is to be thought of as necessary.

III. – THE FIRST SYSTEM IS EASILY REFUTED. – ***1st Argument.*** – Since any member of the series is contingent, the whole series is nothing other than a collection of contingent things. But when a contingent thing is added to another contingent thing the result is a contingent, non-necessary thing, just as a finite number added to another finite number makes a finite number. The whole series, therefore, is contingent. And, in fact, the effect is not greater than its cause. But the whole series is an effect of contingent things. The whole series, therefore, cannot be called necessary, otherwise it would be greater than its cause.

2nd Argument. – It is contradictory for an infinite series of causes to exist, unless there were outside of the series some being that sustained the whole series. For indeed, the last member of the series causes only because it was caused by the preceding one, and the latter moves only because it was produced by another prior to it, and so on; so at last we must reach something that is first and which itself is not caused, otherwise the same question would have to be asked of it. Therefore, there can only be an infinite series of causes if there is another necessary being that subsists outside of the series and which sustains the entire series, whom we call God.

IV. – Uncaused matter. A very ancient error in philosophy, and which in our days has been revived with a new appearance, is that of those who posit an eternal and uncaused matter. This opinion is not only taught by the materialists—Leucippus, Democritus, and Epicurus—but is common to almost all of the pagan philosophers. Plato himself—although he conceded that the world was from God as far as its perfect order and present arrangement—believed nonetheless that the mass of corporeal substances was eternal. It is not clear what was the mind of Aristotle on the issue; it seems nonetheless that he adhered to this error, positing an uncreated and eternal world.[26] Others, however, interpret him as saying that the world was eternal, although created by God.

The modern materialists rehash these very old opinions with an ambitions medley of words. They either posit eternal atoms, or an infinite series of material worlds succeeding each other without beginning and forever without end. Flammarion thinks that, "[t]he universe is in perpetual creation: the geneses of new worlds currently light up the heavens: cemeteries of dead planets circulate in the depths.... a thousand heavens, a thousand earths have just now disappeared in the great night."[27]

All of these believe that matter, its evolution, and its motion, are eternal.

---

[26] Cf. Aristotle, *On the Heavens and the World*, 1.10 (Text 102 in St. Thomas' Commentary).

[27] Flammarion, *Le monde avant la création de l'homme*, p. 15: "L'univers est en création perpétuelle: des genèses de mondes s'allument actuallement dans les cieux: des cimentières de planètes défuntes circulent dans les profondeurs.... mille cieux, mille terres se sont déjà évanouis dans la grande nuit."

V. – CONCLUSION: "The idea of uncreated matter existing eternally is self-contradictory."

1st Argument. – If existence were essential to matter from eternity, then it would also be something essential, without which matter could not exist. But matter can only exist at rest or in motion: it is, of course, impossible to conceive of matter that exists devoid of motion and rest simultaneously. Therefore, it would be essential for matter to be at rest or in motion. But either of these alternatives implies the hypothesis. Therefore, matter existing through itself eternally is contradictory. ***Proof of the first disjunctive part.*** What is essential to a thing is always such, and, therefore, inseparable from the thing. Therefore, in the first hypothesis, rest would be inseparable from matter, and hence matter would necessarily exist and be at rest. Therefore, there would be no activity, or no evolution in the world, but perpetual and immutable sterility. The laws of physics confirm this argument. The *principle of inertia* is very well known: "A body at rest cannot give itself motion." This law cannot be doubted without overthrowing the entire science of [Newtonian] mechanics. For how could the mathematical calculations concerning the navigation of ships, or the trajectory of trains, etc., be any good if the principle of inertia wavered? Therefore, if matter were eternally at rest, it would never be able to move. How, then, could it be now moving? ***The second disjunctive part is established.*** If matter were essentially in motion, it would never be able to come to rest, for what is essentially such, is always and necessarily such.[28] Here again the principle of *inertia* is useful: "A body in motion cannot by itself alone modify its own motion." Therefore, if matter were eternally in motion, it would never modify its own motion, nor ever come to rest.

Whether the adversaries of this thesis choose the first or the second disjunctive part, they are hard pressed by an insuperable difficulty. They are compelled to admit a contradiction, as many of

---

[28] *Sic*, or *tale*, ('such') is a technical Latin scholastic term used to refer to a given quality. Thus, when Hugon says that "what is essentially such, is always and necessarily such," he means that "what is essentially *x*, is always and necessarily *x*," where *x* is a variable that stands for a quality (e.g., "what is essentially human [or mobile, or yellow, etc.], is always and necessarily human [or mobile, or yellow, etc.]"). – ***The Translator.***

them admit. "Yes, if motion has existed from all eternity, one does not conceive that the world has not attained rest and perfection... We touch here on the antinomies of Kant, those depths of the human spirit where *one is jolted from one contradiction to the other.* Once one has arrived there, one must stop."[29]

Du Bois-Reymond, the well-known materialist, says something similar: "Since motion is not essential to matter, our need for causality demands either the eternity of motion, *and then we must renounce understanding anything*, an absolute difficulty for every man of sound spirit, or a *supernatural impulse*, and then one needs to admit miracles, a difficulty full of despair for the positivist."[30]

In sum, if matter exists at rest from eternity, in virtue of the law of inertia it would never give itself motion, and nonetheless it is obvious that matter and the material world are in motion. But if matter were in motion from eternity, in virtue of the same law of inertia, matter would never be able to modify its motion, and nonetheless it is obvious that motion is modified, and that the world is subject to an alternation of motion and rest. Therefore the world does not have movement from itself. Therefore, both motion and existence are given by an unmoved and necessary Being.

The mechanists respond that eternal motion, precisely because it is eternal, does not require a cause. But given that motion is eternal, in no way is the first mover, which is distinct from the world, excluded. For each motion is contingent; this is clear from what was said in no. III above.

Moreover, if matter had been in motion from eternity it would not have been more perfect then than it is now. But now, in virtue of the law

---

[29] RENAN, *Dialogues philosophiques*, p. 146: "Oui, si le mouvement a existé de toute éternité, on ne conçoit pas que le monde n'ait pas atteint le repos et la perfection... Nous touchons ici aux antinomies de Kant, à ces gouffres de l'esprit humain où *l'on est ballotté d'une contradiction à une autre*. Arrivé là, on doit s'arrêter."

[30] *Speech before the Berlin Academy*, July 8, 1880: "Le mouvement n'étant pas essentiel à la matière, le besoin de causalité exige ou l'éternité du mouvement, *et alors il faut renoncer à rien comprendre*, difficulté absolue pour tout homme sain d'esprit, ou une *impulsion surnaturelle*, et alors il faut admettre le miracle, difficulté désespérante pour le positivisme."

of inertia, it can neither give itself motion nor modify its motion. Therefore, it could not have given itself motion from eternity; for this reason, if it were in motion from eternity, it would be moved by another, namely, by the first Mover. The principle of inertia absolutely requires this.

The dynamists object that matter is active, and, thus, gives itself motion. But this is absurd. For even if matter were active, it is still not pure act, but is in itself in the state of potency, in the passage from potency to act, from rest to motion, and from motion to rest. Therefore, since, in virtue of the law of inertia, potency cannot give itself act, nor can matter bring itself from rest to motion or from motion to rest, it always requires a first Mover from which it receives all this.

They reply that matter is living, and that the living can move itself. – But they would have to prove that life belongs essentially to [all] matter. Moreover, although the living can move itself, it has different parts of which some are movers, others are moved, and the mover is in another respect in potency, and thus it requires another thing by which it is first moved. Hence, the axiom, "Everything that moves is moved by another" (*Omne quod movetur ab alio movetur*) is universally true.

Others, like Kant, think that the present world could have arisen from a cloudy mass of primitive matter, through the force of gravity alone, without any extraneous impulse.

But first, the force of gravity has not been proven to belong essentially to matter, or to be in act from eternity. Even if this were proven, there would be no reason why motion or orbits go in one direction rather than another: gravity would be equally everywhere, and hence, there would be no motion.

"If we, like Kant, acknowledge only gravity and the mutual actions of the corpuscles of the nebula, then orbital movements, equally possible in any direction, will be produced effectively in all directions at the same time. Amidst the molecules of this vast nebula, some will be produced to the right, others to the left; but, thus, if you consider the vector rays of all those molecules projected over a given point, those projections—some negative, others positive—will have a sum that is rigorously zero, because they describe opposite directions. That is what [Newtonian] mechanics will require."[31]

2nd Argument. Uncreated matter would be independent in *esse* if it existed by its own power. But an *ens* that is independent in *esse* also acts on its own right, since acting is proportionate to *esse*. Therefore, matter would be independent in acting. Now, it is clear that the action of matter is dependent on many things. For that is called 'independent' which is of its own right, and neither tends to, nor is ordered to, another. But the acting of matter, by the very fact that it is incomplete, tends to something ulterior because it can be perfected and is subject to evolution. Therefore, it lacks independence; therefore, matter is not uncreated.

3rd Argument. Unmade matter would be simultaneously infinite and finite, with regard to both its essence and its quantity, which contradicts the conclusion.

(1) It would be infinite with regard to its essence. Uncreated *ens* is independent in *esse* and in acting, as was said. But an *ens* that is independent in *esse* and in acting is necessarily infinite with regard to essence. Therefore, uncreated matter would be infinite with regard to essence. – And it would also be simultaneously finite. For matter evolves; it is in the process of becoming. But that which evolves and is in the process of becoming is finite with regard to essence, for the infinite, insofar as it contains all perfections in act, cannot evolve. Therefore, matter is finite as far as its essence.

(2) It would be simultaneously finite and infinite with regard to quantity. The quantity would be finite, because the argument showed that quantity that is infinite in act is contradictory.[32] – It would be

---

[31] FAYE, *L'origine du monde*, 2nd Ed., p. 135: "En ne tenant compte, comme Kant, que de l'attraction et des actions mutuelles des corpuscules de la nébuleuse, les mouvements de circulation, possibles également dans les deux sens, se produiront effectivement dans les deux sens à la fois. Parmi les molécules de cette vaste nébuleuse, les unes prendront leur droite, les autres leur gauche; mais alors, si vous considérez les aires décrites par les rayons vecteurs de toutes ces molécules et projetées sur un point quelconque, ces projections les unes positives, les autres négatives, parce qu'elles seront décrites en sens contraire, auront une somme rigureusement nulle. Ainsi le veut la méchanique."

[32] Cf. *Ontology I*, Treatise II, q. 2.

infinite, because an eternal, uncreated, independent quantity does not have a cause that determines it to a given dimension. This is the common position of the Scholastics.

VI. – On the Fortuitous Coinciding of Atoms. Though refuted by these arguments, the opinion of those materialists who posited that the world began from the fortuitous coinciding of atoms still remains. This view is refuted in many ways.

(1) The said atoms would not be in motion, nor would they be at rest, on account of the first argument.

(2) If, therefore, they are not endowed with motion, then *a fortiori* they do not enjoy contrary motion or contrary direction such that they may meet and form one continuous thing.

(3) They would be neither finite nor infinite in number. They would not be infinite, because an infinite number is contradictory.[33] Nor would they be finite, because there would be no cause that could determine atoms to a certain number. Further, such a cause is required, for atoms are of themselves indifferent to such or such a number.

(4) In the world there is a marvelous order by virtue of which similar and dissimilar things are arranged in their respective places. But chance cannot be the cause of so marvelous an order. Therefore, the world did not begin from the fortuitous coinciding of atoms.

***Proof of the Minor*.** Those things which come to be by chance, without a directing cause, happen *per accidens*, rarely, and not always or in a uniform manner. Thus, from the fortuitous coinciding of letters perhaps one phrase could result, but it is altogether impossible that a whole poem or a whole book, in which each letter is arranged in its own order and place, were to be composed without a directing cause. Therefore, it is also impossible for this marvelous poem that is the world and the order of the world to have come out of the fortuitous coincidence of atoms. "Let us suppose that you find ten letters forming the word 'absolutely'. In this case, you no longer hesitate and you affirm without fear of error that the author of this juxtaposition [of letters] knew how to read and wanted to form the English word that you read... Calculations

[33] Cf. *Ontology I,* Treatise II, q. 2.

show that the odds are 3,628,800 to 1 in favor of your conclusion... It is enough to increase by a little the number of letters that form the remarkable arrangement to pass from an ordinary probability to a practical certitude."[34] But if we are dealing, not with ten or a hundred letters, but with innumerable words that make up a poem, it is certain that there is no chance involved whatsoever. Since the order of the world is more difficult and entangled than the arrangement of all the letters of a poem, it is most certain that the world is not the effect of chance.

It cannot be said that all possible arrangements have already been accomplished, supposing that time is infinite. For, in order for all hypotheses to be verified, at least the law of probability is to be admitted. But where there is only chance, no law is conceived. Therefore, these hypotheses could not have been accomplished.

Moreover, chance arrangements, insofar as they are unstable, do not take place in the future as they did in the past, they do not take place today as they did yesterday, nor will they take place tomorrow as they did today. Therefore, supposing that those arrangements once constituted the world, the world would not remain for long, but other combinations or arrangements would arise, which would dissociate the atoms and entirely destroy the world.

For unless there is a law or cause that contains the atoms, there would not be a reason for them to remain in one and the same combination. For this reason, we cannot be certain that the present arrangement will not be dissolved tomorrow or that the Sun will rise tomorrow. If one were to respond that this cannot occur because nature proceeds through determinate phases, already a certain law is being introduced and chance is then eliminated.

---

[34] D. POISSON, *Recherches sur la probabilité des jugements en matière criminelle et en matière civile*, in P. CARBONELLE: *Les Confins de la science*, T. II, IX: "Supposons que vous trouviez 10 lettres formant le mot 'absolument'. Ici, vous n'hésitez plus et vous affirmez sans crainte d'erreur que l'auteur de cette juxtaposition savait lire et a voulu former le mot français que vous lisez... Le calcul montre qu'il y a 3628800 à parier contre 1 en faveur de votre conclusion... Il a suffi d'augmenter un peu le nombre des lettres qui forment l'arrangement remarquable, pour passer d'une probabilité ordinaire à une certitude pratique."

VII. – The System of Active Evolution. It is defended by Strauss, Buchner, Haeckel, and others, although not in the same manner. According to them there is neither matter without powers, nor powers without matter—one cannot be conceived without the other. From this principle, they argue thus: Powers cannot be without matter, nor can they be prior to matter. But, if matter were produced, the power which would give existence to matter would be prior to matter, which is impossible. Therefore, matter is not produced. Further, matter cannot exist without powers. But, if powers were produced, then the matter which generated them would be prior to the powers, which is also impossible. Therefore, it is impossible for powers to have been produced. Matter and powers, then, are uncreated, eternal, and indelible. Matter with powers, or, in other words, material powers, are compelled by an insuperable necessity to an eternal series of evolutions. Material powers contained in the beginning the seeds of all things and, by means of conditions that helped each other evolve, they obtained vegetative life, later sensitive life, and finally rational life. The human soul, therefore, is the last result of material powers.

Evolutionism is proposed in a new form by the Darwinists; we shall discuss these things below, where we treat of the origin of species.

VIII. – The Theory of Active Evolution Suffers from Multiple Absurdities. (1) *Power* signifies an active principle. But the concept itself of active principle does not in any way imply matter; on the contrary, the more elevated it is above matter and the more independent it is from matter, the more efficacious it is. Therefore, it is false to say that it cannot be conceived without matter. (2) Matter that is not produced is self-contradictory, as we showed. (3) Matter that acts by a blind impulse cannot generate such exceptional harmony, such ornate variety, things being arranged in such a way that each has its own end, and means perfectly proportionate to that end. For the cause of order cannot be chance, or blind necessity, as was proved in no. VI.

IX. – The system of hylozoism. (*hyle*-, "matter"; *zoe*, "life"). This is the error of those who posit one soul in matter that pervades the whole world, and that informs everything, and which mixes itself with all bodies, and which from eternity evolves itself forever by fixed and invariable laws. We sufficiently refuted this error by refuting Monism (in the preceding Article).

Peter Abelard claimed something similar, when he said "that the Holy Ghost is the soul of the world." This proposition was condemned by the Council of Sens and by Innocent II.

X. – The System of Logical Evolution. According to the Hegelian opinion, the absolute, the ego, the idea, evolves by logical necessity and the world comes to be, almost by that necessity whereby a conclusion is derived from the premises. But, on the contrary, in logical evolution, nothing comes to be *per accidens* and contingently: rather, a logical conclusion is derived from premises *per se* and necessarily.

But in the world, many things come to be contingently and *per accidens*: for it is clear that often causes are impeded in their effects, or that they produce a chance effect. Therefore, logical evolution in the world is contradictory. Renan restored the Hegelian opinion with a new appearance, by calling God the ideal category, "the transcendental summary of our sensible needs, the category of the ideal."[35] But this ideal is in no way capable of making the world, for it is an idea posterior to reality and posterior to the intellect.

Taine tried to explain the origin of things thus: "At the highest summit of things the eternal axiom is pronounced, and the prolonged effect of this creative formula composes the immensity of the universe through its endless undulations."[36] Further, this axiom is either a certain reality or a mere abstraction. If a reality, then the intellect presupposes an eternal being that brings forth the eternal axiom; for an idea or axiom does not produce an intellect, but an intellect produces an idea or an axiom. If a mere abstraction, how can it give existence to sensible realities and to the concrete world?

---

[35] *Liberté de penser*, t. 4, p. 348: "Le résumé transcendental de nos besoins sensibles, la catégorie de l'idéal."

[36] *Philosophes français*, p. 364: "Au suprême sommet des choses se prononce l'axiome éternel, et le retentissment prolongé de cette formule créatrice compose par ses ondulations inépuisables l'immensité de l'univers."

XI. – What Kind of Necessity is Found in the World. Certain scientists try to take away contingency from the world by means of this argument: That is necessary in which many immutable and indestructible things are found. But in the world, there are many such things: thus the laws of nature are wholly immutable; the amount of energy in the world is constant and invariable; matter is indestructible, for despite changes occurring, the same weight remains. Therefore, the world is a necessary being.

Solution: All these are evidence for a certain hypothetical necessity. Supposing that the world has been endowed with such an order, it is not ruled by chance, but certain things in it must be, given that order; but from that fact one cannot conclude an absolute necessity, but only a conditional necessity. Therefore, there is a reply to each objection:

(1) The laws of nature are necessary in such a way that nonetheless they could be otherwise, or even not be at all, and such that even now exceptions occur. See below, Treatise III, Question II, Article I.

(2) The Law of Conservation of Energy obtains only in the mechanical and physico-chemical order, but not at all in the vital and spiritual orders, for it is certain that the sum of thoughts, of free choices, of intellectual energy, etc., varies. Therefore that law has limits. Now, that which is restricted by limits does not possess absolute necessity. But in the physico-chemical order, although the sum of energy is constant, nonetheless, transformed energy reappears in an inferior way: *the degradation of energy*; and in it potency and act are always observed. The state of potency and act cannot be ascribed to an *ens* that is absolutely necessary. For this reason, from the Law of Conservation of Energy one cannot argue that the world is absolutely necessary; rather, from it one can prove that there exists a first Mover distinct from the world who reduces energy from potency to act.

(3) All Scholastics acknowledge the indestructibility of prime matter; but they do not admit that matter is absolutely necessary. Since in the world nothing further is created or annihilated, matter necessarily is the common subject which every generation presupposes and which

every corruption leaves behind. And so, matter is indifferent to all forms that are to be successively assumed; this indifference and indeterminacy indicates potency. For this reason, matter does not require—it in fact excludes—necessity; and, even if it in fact always subsists, absolutely speaking, nonetheless, it is reducible to nothingness.[37]

XII. – The Last System is Pantheism.

Those things that were said of monism also serve to refute pantheism. But because this most recurrent error remains unrefuted and takes many forms, it is necessary for us to treat of pantheism more fully.

[37] On the unity or origin of the world, the following authors may be consulted: LEPIDI, *Cosmol.*; PESCH, *Phil. Nat.*; DUILHÉ DE SAINT PROJET, *Apologie scientifique*; MONSABRÉ, *Carêmes de 1873 et 1875*; FAYE, *L'Origine du monde*; FARGES, *L'Idée de Dieu*; SERTILLANGES, *Les sources de la croyance en Dieu*; and the authors cited below concerning creation.

THIRD ARTICLE

ON PANTHEISM

I. – The Notion of Pantheism. The term 'pantheism' was coined during the time of Spinoza, but the theory to which it refers is very old. This error, taken in general, can be defined as: *The opinion that asserts that the world and all the things that are in the world are identical with God, with an identity of both substance and existence.* Therefore, it is characteristic of this evil system that it posits consubstantiality between nature and God, between the finite and the infinite. But from the identity of substance is inferred an identity of existence. Therefore, there is a single existence, or a single Substance, and the diverse beings of the world are only its modifications or evolutions. Pantheism is not to be understood as saying that God is absorbed by the world, the infinite by the finite, or vice versa, as some seem to understand it; but such that God and nature, infinite and finite, *coexist simultaneously in one consubstantiality*, as two diverse and inseparable aspects of the universal Substance and existence. The pantheists, therefore, assert two things: (1) that there is a single Substance and a single existence in the world; (2) that in that universal identity there is evolution without limit, in such a way that through its indefinite unfolding all things are made. For this reason, that Substance is most indeterminate and most potential.

II. – Defenders of Pantheism. The old Monists, namely, the Eleatics and the Stoics, of which we treated in Article I, were pantheists. Indian emanationism is reducible to pantheism. The Neo-Platonists also taught a certain emanationism that logically contained pantheism. In the middle ages, Scotus Erigena claimed that the nothingness out of which created things were made is the divine nature, and that the divine nature alone is in all things. Almaric of Bena claimed that God is the formal principle of all things; David of Dinant most foolishly said that God was prime matter.

– At the beginning of the Seventeenth Century, Giordano Bruno, a monk turned heretic and apostate, revived the pantheism of the Alexandrians. In his opinion, God is the *primitive monad and the absolute being*, from which all things emanate and which constitutes their essence. The world is animated by God and can be called the *holy, sacred, and venerable living thing*. Further, in that Substance there is infinite potency, or "the potency of potencies," by means of which all things can be made, and an infinite actuality by means of which all things are in act; hence, this Substance can be called "the form of forms."

Insofar as it is in potency, this Substance constitutes the matter of the world; insofar as it is actuality, it is the soul of the world, or "the life of all lives, the soul of all souls." It is whole in the entire world, and whole in each being, and for this reason it evolves first into the extended matter of bodies, and then it becomes the unextended matter of spiritual things.

Bruno prepared the way for Spinoza. He developed and refined the pantheistic system scientifically. He begins with the Cartesian definition of 'substance': *That which does not depend on any other thing to exist*. From this, he argues thus:

(1) The Substance is its own cause. For if it were caused [by something else], it would depend on something else and thus would fall away from the *ratio* of 'substance'.

(2) It is infinite. For if it were limited by something, it would be dependent and again it would lose the *ratio* of 'substance'.

(3) It is one. Because, if there were other substances, it would be limited by them and thus it would become dependent, which would be to fall short of the *ratio* of 'substance'.

There exists, therefore, a single Substance, a single cause that intrinsically constitutes the reality of all beings; it enjoys an infinity of attributes, which are chiefly manifested in extension, by the *ratio* of which body comes to be, and in thought, by the *ratio* of which spirit comes to be. Extension and thought are infinite, in the sense that thought completes whatever can be a thought, and extension whatever can be extension. Further, this single Substance and Cause, of which the diverse beings are only modifications, is God Himself. Nonetheless, it is designated with various names. If God is considered the efficient and constitutive cause of things, He is called *Natura naturans* [or 'active nature'; lit., "naturing nature"); but if He is seen as a substance that is modified and diversified in the manifestations that occur in the world, He is called *Natura naturata* (or 'passive nature'; lit., "natured nature"). Further, God is thought of as free, because He is compelled by no one else, but nonetheless he acts with a necessity of nature; all others, however, are compelled, because they are impelled to act by another.

III. – Idealist Pantheism. – Fichte. The Pantheism that has been hitherto exposed is called realist pantheism; but now we must arrive at the ideal form of the German transcendentalists. Kant did not teach pantheism; but from his principles others deduced a pantheistic idealism; especially Fichte. Here is a summary of his system: There is no reality beyond the thinking subject, *the pure Ego*. The first principle, and the most certain thing in every science, is this: *Ego = Ego*, *I am myself*. By virtue of this principle, the *Ego* posits itself, and this is the supreme and pure activity, to posit oneself. But at the same time that the *Ego* posits itself, it exists. This is the first moment in the evolution of the *Ego*. This primitive affirmation of the *Ego* is the *thesis*. Then the *Ego* conceives itself as denied, or as the *Non-Ego*. Further, this *Non-Ego* is that by which the *Ego* becomes objective, and thereupon is the external world: this is the *Antithesis*. But by the *Ego* positing itself before the finite *Non-Ego*, it restricts itself in some way, and thus it becomes determinate and finite; and, given this limitation, a certain union is completed between the *Ego* and the *Non-Ego*: this is the *Synthesis*.

There are, therefore, three moments in the evolution of the *Ego*: the primitive Affirmation of the *Ego*, the *Thesis*; the Negation of the *Ego*, or the affirmation of the *Non-Ego*, the *Antithesis*; and the mutual union of the *Ego* and the *Non-Ego*, the *Synthesis*. But there is no reality beyond the *Ego* positing itself as Affirmation, as Negation, as Limitation.

IV. – Schelling noticed that Fichte's claim that the *Non-Ego* is derived from the *Ego* is gratuitous; and by a similar reasoning it can be said that the *Ego* arises from the *Non-Ego*. It is necessary, therefore, for us to find a principle in which all opposition between the *Ego* and the *Non-Ego* are removed, namely, an indifferent principle in which all things are identified, and which is called *The Absolute*. In it all things preexist without any opposition, as being identical: namely, subject and object, finite and infinite, universal and singular, real and ideal, *Ego* and *Non-Ego*, spirit and nature. The Absolute is the substrate of all things in which contraries disappear; before its evolution nothing comes from those contraries, but it is able to become all things. It can be called "the indifference of indifferent things, or the principle whose form is indifference, and whose essence is the universal identity."

Further, all things are one in it; hence, the duality that we think exists between subject and object, finite and infinite, is only fictitious. The Absolute is God Himself: in the state of principle He is an *implicit* God; but in the state of end He is an *explicit* God. But God becomes the world through a certain *leap*, or through a certain self-lessening by which He is constituted in an inferior degree.

V. – Hegel claimed that the only reality is the *Idea*. The *Idea* is the principle, the essence, and the terminus of all reality. The Idea is *becoming* itself, and it evolves in three phases: in pure *concepts*, or in the ideal order; in the *order of nature*, or that of the external world; and in the *order of the mind*. To this threefold evolution corresponds the threefold division of Philosophy: *Logic*, *Philosophy of Nature*, *Philosophy of Mind*. The three aforesaid parts of philosophy are diverse determinations of the same Idea, and they rest, not on the principle of contradiction, but on this principle: *Whatever is rational is real; whatever is real is rational*. Hegel rejected the principle of contradiction. Further, the Idea proceeds through three moments: (1) *The Thesis*, or positing, or the Idea in potency, in a state of involution; (2) *The*

Antithesis, negation, or the Idea in a state of evolution, which constitutes the world; (3) *The Synthesis*, the negation of the negation with a positive effect, or the Idea reversed upon itself; this reversion of the Idea upon itself makes the Mind.

– Diverse opinions having been exposed, it is clear that all forms of pantheism are reducible to the generic definition that we offered at the beginning, namely, that the world and God, the finite and the infinite, coincide with an identity of substance and existence.

VI. – CONCLUSION: "Pantheism, whether it be considered in general or under the successive forms that it assumes, is altogether absurd."

Proof. Any opinion that denies common sense, the principle of contradiction, the principle of causality, the existence of God, human free will, and morality, is full of absurdities. But pantheism does away with all these things. *Therefore.*

[***The minor is proved in parts.***]

(1) ***Pantheism denies common sense.*** For common sense bears witness to the fact that we have our own individuality, that our own actions are complete and have a terminus in themselves. But, if all the beings of the world are one Substance with God, all individuality is annihilated; for, of course, an individual is a substance that is undivided in itself and divided from any other. Equally, no action that is complete in itself will be elicited; hence, our volition will not have a terminus nor will be *sui juris*, but something partial and suspended that is ordered to the whole Universal Substance. Pantheism, therefore, clashes with common sense.

(2) ***It denies the principle of contradiction.*** The first argument with which we refuted Monism in Article I is useful here. It may be otherwise proposed: He who claims that the same subject is simultaneously imperfect and perfect in act rejects the principle of contradiction. But pantheism claims that the same subject is simultaneously perfect and imperfect in act. *Therefore.* ***Proof of the Minor.***[38] Pantheism claims that the subject is actually perfect because

---

[38] That is, the minor of the argument under point (2). – ***The Translator.***

this subject is said to be infinite, and the infinite does not lack any perfection. And it claims that it is simultaneously imperfect in act for, according to the pantheists, this subject undergoes an endless evolution; but nothing evolves unless it is because it needs a further perfection and, therefore, because it is imperfect in act.

Further, to conjoin in one subject many formal beings (*plura esse formalia*) that are contradictory, contrary, or privative opposites is to throw out the principle of contradiction. But pantheists conjoin in one subject many formal beings (*plura esse formalia*) that formally are opposites. *Therefore.* ***Proof of the Minor***. The pantheists conjoin in one subject all the things that are in the world, namely, all the determinations that we observe in nature, whether generic or specific or individual. Further, these determinations are such that the formal being (*esse formale*) of one is opposed to the formal being (*esse formale*) of the other, either contradictorily, or contrarily, or privatively, as is apparent to anyone who considers the matter: for example, living and non-living, man and non-man, body and spirit, virtue and vice. The pantheists, therefore, conjoin in one subject many formal beings (*plura esse formalia*) that are contradictory, contrary, or privative opposites.

(3) ***It denies the principle of causality***. For the principle of causality also states that potency is reducible to act through some act and, thus, that act is absolutely and *simpliciter* prior to potency. But in the doctrine of pantheism, potency is absolutely and *simpliciter* prior to act. Pantheism, therefore, does away with the principle of causality. ***Explanation of the Major***. In this consists the principle of causality: that every effect has a cause, or that nothing can give itself that which it in no way has. But, if potency is itself reducible to act, it would give itself something that it in no way has. *Therefore.* For potency signifies only the capacity or aptitude for something; but the capacity for some thing in no way implies having that thing in act, otherwise it would not be called a mere capacity. Potency in itself, therefore, does not in any way imply act, and thus if it could move itself through itself, it would give itself something that it in no way has, which is to discard the principle of causality. And so it is necessary that it be reducible to act through something that already has act, and so this would be prior, at least in nature. Therefore, the major premise is established: The principle of causality requires that act is *simpliciter* prior to potency. ***Proof of the Minor***. Since the [supposed] universal Substance is one, no act can be conceived beyond it; hence, if it is in potency before it is in act, one is to conclude that potency is *simpliciter* prior to act. But said Substance is in potency before it is in act. Therefore, in the doctrine of pantheism, potency is *simpliciter* prior to act. ***Explanation of the subsumed minor***.[39] The Universal Substance is conceived in the beginning as something that evolves and is capable of receiving an infinite number of determinations. In the beginning, therefore, there would have to be something potential and it would be perfectly in act to the extent that it receives the last form of evolution.

It cannot be replied that these determinations are not additions of new *esse*, but only new manifestations of infinite *esse*. For the pantheistic notion of *esse* is that of some sort of purely universal *esse* that is distinguished into genus, species, and individuals through its evolution. Now, the determination of being into genera, species, and individuals is not a mere manifestation of universal *ens*, but a new perfection, and the positive addition of new *esse*; in the individual there are, of course, real and positive perfections that are not included in the concept of universal *ens*. The evolution of pantheistic being, therefore, is not conceived as a mere manifestation, but as a true acquisition of perfection. It remains, therefore, that in that system potency is prior to act and that it gives itself an act that it does not in any way have; this is to do away with the principle of causality.

(4) ***It denies the existence of God***. Pantheists claim that they reject atheism, but in reality they are nothing other than atheists. If God exists, He is to be conceived as the most perfect being, containing within Himself all perfections, and all plenitude of *esse*, as a pure act: for men universally understand God to be the being than which nothing greater can be thought. But the Universal Substance that the pantheists construe is impure act, a potential and determinable *ens*, lacking innumerable

[39] The "subsumed minor" is the minor premise of a subordinate argument, in this case the argument immediately above (i.e., of the "proof of the minor"). – ***The Translator.***

perfections, since it is subject to innumerable evolutions. To claim, therefore, that such a Substance is God, is to deny the existence of the true and glorious God, to whom praise and honor is due world without end.

(5) *It denies the freedom of the will.* The freedom of the will requires individuality and personhood. Pantheism eliminates both of these by positing a single impersonal substance. Again, freedom, which requires the indifference of the will, presupposes that the determination of the will is contingent, that it can be given and withheld. But, if we admit pantheism, no volition is contingent, but rather the universal substance is compelled by a necessary and inevitable evolution.

(6) *It denies morality*. This is a corollary of the preceding. God and human freedom having been removed, all the foundations of religion and morality fall apart. Further, if all the beings of the world are somehow divine, any action of theirs is, for the same reason, to be called divine: theft, homicide, blasphemy will be divine and holy actions meritorious!

VII. – Refutation of Pantheistic Systems in Particular. The foregoing are the impossibilities that follow from pantheism in general; but beyond them, there are particular absurdities that assail each particular form of pantheism. Thus, those who posit a universal soul in the world, or that the world is a living thing, try to claim that all things are alive, and so confuse the mineral kingdom with the superior kingdoms.

Spinoza holds contradictory claims, affirming that substance is simultaneously infinite and extended; of course, the infinite and the extended are mutually exclusive. Further, he perverted the notion of substance. For substance is not that which *needs no cause* to exist, but only an essence that *does not need a subject* in which to inhere.

Fichte imagined monstrous things. Why can the *Ego* posit itself? Fichte affirms this but in no way proves it. How is the negation of the *Ego* able to produce the external world? Why cannot the *Ego* proceed from the *Non-Ego*? – Further, he falls into a contradiction when he asserts that the *Ego* is infinite and nonetheless that it is limited, which limitation is called the "synthesis."

Schelling claimed that contraries and contradictories are the same, if the absolute is simultaneously finite and infinite, real and ideal, etc. Moreover, he asserted gratuitously and without any apparent proof that the absolute exists of itself, and that nothing enjoys reality beyond it. Absurdly, Hegel denies the principle of contradiction. Further, it is senseless to claim that an idea is the first reality, since an idea is only a representation, an image, and it is false to posit an image before the thing itself that it represents; for the whole being of the image consists in the fact that it stands for the thing.

VIII. – On Certain Other Forms of Pantheism. – Krause introduced psychological pantheism. Our *Ego*, or conscious awareness, is identified with the universal conscience; hence, from our *ego* is to be taken the principle of all sciences. The eternal *ens*, or God, is manifested in us and through us, but other external beings are only phantasms or modifications by which the infinite *ens* is made known. Krause also wants to retain the existence of a personal God. The world is in a certain way part of God. God and the world are together in one essence, in such a way, however, that God is that essence entirely, the world not entirely.

It is clear that, although conscious awareness is a criterion in its own order, it is not the principle and source of all cognition. Otherwise, if God is of one essence with the world, His personal *esse* is lost and He is no longer pure act; and then all the absurdities of pantheism in general resurface.

Schopenhauer put forth the system of *panthelism*, that is, of a single, perverse, and most terrible will. This opinion is a delusion, and it would be more appropriately called 'pan-satanism', rather than 'pan-theism'.

Hartmann invented the system of the *unconscious being*. The principle of things is the identity of will and cognition, which identity is fulfilled by representation. Now, it is necessary that a representation lack a consciousness, since consciousness is an imperfection and it involves a calamity. For this reason, the principle of things is unconscious. The world is only the appearance of this volition; and whatever is, is one. Man is constituted by the *Unconscious*, and he has no immortality beyond that which he obtains from his union with the *Unconscious*, as a phenomenon or manifestation of the divine Substance. The *Unconscious* evolves in three phases: in paganism, in Christianity, in modern times, and these are the three stages of the illusion. This theory is insanity, greatly injurious to the true and glorious God, who, containing all perfections in act, cannot lack a conscience, or evolve into images and appearances.

We we have referred to the opinions of Renan and Taine in the preceding Article (no. X).

IX. – PANENTHEISM. – Pan*en*theism is a theory that claims that God is *in* all things, although all things do not add up to God. This system was introduced by Krause, and was then proposed in a new form by P. Janet.[40] Janet teaches that God is in all things, although He is not all things. – If God is in all things only as an efficient cause, infusing and conserving *esse*, then the theory is true; but, if God is in all things constituting the *esse* of all things, then God becomes all things. For that which formally constitutes the *esse* of all things mixes its being with all things, and consequently has one *esse* with all things. Panentheism, therefore, destroys the being and personhood of the true God.

X. –THEOSOPHICAL PANTHEISM. H. P. Blavatsky explains it thus: "Theosophy is the divine science.... We say that the divine spark within man is one and identical in essence with the universal Spirit; and that, consequently, our spiritual Ego is in reality omniscient, but that the obstacles of matter prevent it from manifesting its knowledge... According to our teachings, Spirit and Matter are identical; Spirit contains Matter in a latent state, and Matter is nothing but crystallized Spirit, as ice is solidified vapor. Humanity is absolutely of one and the same essence, and this essence (which we call God in nature) is one, infinite, uncreated, eternal... The root of all nature, both objective and subjective, and of that which is found in the Universe, both visible and

[40] *Revue des Deux Mondes,* June 1, 1885.

invisible, is, has been, and will always be one single absolute essence, from which all proceeds and to which everything returns.... Our deity... is found everywhere: in every atom of the Cosmos, visible and invisible.... We reject the idea of one, personal, or extracosmic, God."[41]

These delusions are injurious not only to God but also to common sense; the arguments exposed so far abundantly suffice to blow them away.

XI. – Difficulties Resolved.

Objection: If there were other substances besides God, God would be limited by them. But God, insofar as He is infinite, cannot in any way be limited. Therefore, there are no other substances besides God. ***Reply***: *I distinguish the major.* That, if there were other substances besides God of His same order and perfection, then God would be limited by them, I concede; but that God would be limited if there were other substances of an inferior order that were dependent on God, I deny. *And I deny the consequence.*[42]

The infinite is not to be defined as, "That outside of which there is nothing," but as "That outside of which there is nothing of the same perfection." Now, if outside of the infinite something is found that is dependent on it, the infinite is not thereby limited, but rather the finite by the infinite. In fact, the limitlessness of the infinite is manifested by the fact that all other things are limited by and depend on it.

You will insist:[43] (1) If other substances are really distinct from God, then some reality is found in them that is not in God. But if God is not in every reality, then God is not infinite. *Therefore.* ***Reply***: *I distinguish the major.* That it is necessary for those substances to have a reality that is not in God according to a material and formal *esse*, I

[41] Helena Blavatsky, *The Key of Theosophy*, New York: W.Q. Judge, 1889.

[42] The "consequence" is simply the conclusion of the objection. From now on, Hugon will use this term for the conclusions of objections, and will reserve the term "conclusion" to the main theses that he defends. – ***The Translator.***

[43] The "You will insist" (*instabis*) is not a new objection, but a revised version of the same objection, after it has been provisionally refuted. In this section (no. XI), there is one objection and five *instabis*. The use of the second person ("you") is customary in Scholastic disputations; Hugon is anticipating the reader's objections as if it were a live disputation. – ***The Translator.***

concede; but that they have a reality that is not found in God according to some more perfect *ratio*, I deny. *I contradistinguish the minor*. That God would lack infinity if each reality were in God neither according to formal and material *esse* nor in a more eminent manner, I concede; but that God would be finite if every reality is in God, not according to formal and material *esse*, but according to a more eminent *esse*, I deny. *And I deny the consequence.*

Every reality that is in the world is to be found in the infinite according to a more perfect or equally perfect being; but it pertains in no way to the notion of the infinite that every perfection be ascribed to it formally. For there are certain realities, such as *bodies*, *extension*, etc., that bear an imperfection that is attached to their formal concept, which imply a limit. If, therefore, such a reality belonged to God according to material and formal *esse*, then God would be said to be formally imperfect and limited. The *ratio* of the infinite, therefore, does not require, but rather, completely excludes, the possibility that every reality be formally in God. For this reason, those who assert that God is formally all things necessarily deny that He is pure act and infinite, and claim that he is a potential, determinable being capable of evolving. But the *ratio* of pure act in God is saved if it is established that whatever there is of reality and perfection in beings, for example in bodies, is found in God virtually, that is, in a more sublime and eminent mode.

You will insist: (2) The infinite God has plenitude of *esse*. But plenitude of being exhausts all *esse*. God, therefore, exhausts all *esse* and, thus, nothing other than God can be conceived to be (*esse*). – ***Reply:*** *I distinguish the major*. That God has plenitude of *esse* in the sense that all perfections are in God, and that all *esse* is either God or something that participates in God and is dependent upon God, I concede; but that He has plenitude of being in the sense that all *esse* is God, and that God is formally all *esse*, I deny. In order for God to gather within Himself the plenitude of *esse*, it is not necessary that He be all *esse*; rather, if He were formally all *esse*, by that very fact He would lose the plenitude of *esse*, for then He would become formally extended, potential, and capable of evolving. Therefore, the sense of this dictum, "God has plenitude of *esse*," is that God gathers within Himself all perfections, and all the *esse* of creatures is derived from God and is found in God according to a more sublime mode. *I distinguish the conclusion.* That, beyond God, no other *esse* is conceived that is not in God or from God, I concede; but that, beyond God, no other *esse* is conceived that is not God himself, I deny.

You will insist: (3) God is said to be in things. But that which is in a thing is of one substance with the thing itself. God, therefore, is of one substance with the thing. ***Reply:*** *I distinguish the major.* That God is in things as the *cause* of things, I concede; but that God is in things as *something belonging to things*, I deny. *I contradistinguish the minor.* That what is in a thing *as something belonging to it* is of one substance with the thing, I concede; that what is in a thing only *as its efficient cause* is of one substance with the thing, I deny. *And I deny the consequence.*

The pantheists fall into a great confusion, "not understanding that God is not in things as *something belonging to them*, but as the cause of things, which is in no way absent from the thing. For we do not say that a form is in a body in the same way that a sailor is in a ship."[44]

You will insist: (4) The infinite together with many other substances would form something greater than the infinite alone. But there can be nothing greater than the infinite alone. Therefore, there cannot be many other substances together with the infinite. – ***Reply:*** *I distinguish the major.* That they would form something greater than the infinite alone if they had a univocal *ratio* with the infinite, and if they were independent from it, I concede; but that they would form something greater than the infinite alone if they were only analogous participations of the infinite itself, I deny. *And I deny the consequence.*

If many substances coincided with God in a univocal *ratio* and were independent from God, then the whole *ratio* of substance would be saved not in God alone, but both in God and in the other substances; the plenitude of *esse* would not be altogether found in God, but in God and in other substances together; therefore, all other substances together with

---

[44] St. Thomas, *Summa contra gentiles*, 1.26: "[N]on intelligentes quod non sic [Deus] est in rebus quasi aliquid rei, sed sicut rei causa quae nullo modo suo effectui deest. Non enim similiter dicimus esse formam in corpore, et nautam in navi."

God would make up something greater, or some perfection greater than God.

But if other substances are only analogous participations of the divine substance, they cannot make something intensively greater. For something intensively greater to arise, it is necessary that the other substances add a new reality that is not found in God. But whatever there is of reality and perfection in other substances already preexists in God in a more eminent mode. The other substances together with God, therefore, cannot form something intensively greater. Therefore, a greater perfection does not result from the creation of new substances, nor is the plenitude of *esse* increased.

– ***You will insist:*** (5) The *ratio* of *ens*, was extended only to God, but now through creation it is extended to new substances. The *ratio* of being, therefore, becomes greater. ***Reply***: *I distinguish the antecedent*. That the *ratio* of *ens* is extended also to other substances, in the same sense that it belongs to God, I deny; but that it is extended to other substances in an analogous sense, I concede. For the *ratio* of *ens* belongs to God *per se* and *a se*, but to other substances *secundum quid* and with respect to God, in which they preexist according to a more eminent *esse*.

That, therefore, the *ratio* of *ens* becomes greater *extensively* and *extrinsically*, I concede; that it becomes greater *intensively*, I deny. It can be said to be greater extensively, because it is predicated of many others besides God; but by no means intensively, for the *ratio* of *ens* does not become fuller by being attributed to other substances; but rather, the whole and entire ratio of *ens*, and the whole and entire plenitude of *esse*, is there even if it is predicated of God alone.[45]

[45] The following authors can be consulted: ST. THOMAS, *Summa theologiae* Ia, qq. 3, 4, and 45; *Summa contra gentiles* I.26; BOSSUET, *Elévations sur les Mystères*; GRATRY, *Connaissance de Dieu*; SAISSET, *Introduction aux oeuvres de Spinoza*, *Essais de Philosophie religieuse*; Maret, *Essai sur le Panthéisme*; MONSABRÉ, *Carême de 1873*, *La personalité de Dieu*; PESCH, *Phil. Nat.*: DE SAN, *Cosmol.*; Farges, *L'Idée de Dieu, IIe partie*: L. JANSSENS, *Summ. Theol.* vol. 4, sect. 2, append. 2 in which Nietzsche's theories are beautifully refuted.

SECOND QUESTION

On the true origin of the world

Composition and contingency, which are essential characteristics of the world, are evidence that the world is by another (*ab alio*). Now we ask who the author of the world is and how the world was produced.

FIRST ARTICLE

WHO IS THE AUTHOR OF THE WORLD

I. – CONCLUSION: "The Author of the world must be a Necessary *Ens*, distinct from the world, endowed with intellect and will, pre-containing in Himself all the perfections of things."

This is a corollary of the preceding. We showed that the *ratio* of the existence of the world does not belong to the world in virtue of itself. But that which does not contain in itself the *ratio* of its existence, borrows its *esse* by that which in itself possesses the *ratio* of its own existence, otherwise there would be a process *ad infinitum*. The Author of the world, therefore, is an *ens* that gathers in itself the *ratio* of its own existence. But an *ens* that possesses the *ratio* of its own existence in virtue of itself is a Necessary Being.

The author of the world, therefore, is:

(1) *A Necessary* Ens.

(2) *A Being* Ens *distinct from the world.* This is already clear. The world is contingent, and the Author of the world is a Necessary *Ens*. The Author of the world, therefore, is distinct from the world. Similarly, it is clear from the refutation of pantheism. But it is further proved through a very profound argument that the Angelic Doctor developed:

> Those things that are in the world are diverse and nonetheless exhibit something common: for they share either in common the nature of their species or in the nature of their genus, or at least the *ratio* of *esse*. But they cannot both be diverse and have something common unless they are from a common cause that is really distinct from them. All the things that are in the world, therefore, are from a common cause that is distinct from them; it follows that the cause of the world is extrinsic to the world.[46]

Proof of the Minor. Things that are distinct do not have in common that which is proper to each; this is clear, for two things are distinguished from each other precisely according to that which is proper to each. It is necessary, therefore, to reduce that which they have in common to some single cause, such that one is the cause of the others or that there is one cause common to all. But one cannot be the first cause of the others, because, since each *ens* in the world is contingent, "none of them is sufficient to give to others *esse simpliciter*." It is necessary, therefore, that all things have a common First Cause, which is the cause of the *esse* of all things, giving *esse simpliciter* to all. Further, that Universal First Cause is extrinsic to the world because, as was just said, no *ens* in the world is sufficient to give to others their *esse simpliciter*. That Universal Cause, therefore, is extrinsic to, or distinct from, the world.

(3) ***A Being (Ens) that is endowed with intellect and will***. Order requires an agent endowed with intellect and will. But a marvelous order shines forth in the world. The world, therefore, requires an Author that is endowed with intellect and will. ***Proof of the Major***. Order requires someone wise, because "it belongs to the wise to order" (*sapientis est ordinare*), that is, to dispose things to their proper end. And "the wise" here means an intelligent and free agent. Further, since

---

[46] ST. THOMAS, *De potentia*, q. 3, a. 6: "Ea quae sunt in mundo diversa sunt et tamen exhibent aliquid commune: communicant enim vel in natura speciei vel in natura generis, vel saltem in ratione essendi. At non possunt esse diversa et habere aliquid commune nisi sint ab una causa communi realiter ab ipsis discreta. Ergo omnia quae sunt in mundo sunt ab una causa communi ab ipsis distincta; subindeque mundi causa est mundo extrinseca."

order is the right disposition of things to their proper end, it presupposes an author that knows the relationship between the means and the end, and thus someone intelligent, who wants to dispose the means to the end, and thus, endowed with a will. ***Proof of the Minor***. A twofold order shines forth in the world, namely a *dynamic* order and a *teleological* order. (1) First, there is a *dynamic*, or causal, order. For we see that all beings act upon each other, such that there are active and passive things in the world. Superior beings are disposed in such a way that they know and change inferior things, and inferior beings are so disposed that they are known and changed by the superior beings. Celestial bodies act upon terrestrial bodies, and terrestrial bodies act upon each other, and by a mutual action they constitute the worldly order and the cosmic harmony. (2) Second, there is a *teleological* order, or an order of ends. For the things in the mineral kingdom are for the sake of things in the vegetative kingdom, and these for the sake of things in the animal kingdom, and finally things in the animal kingdom for the sake of man.

Further, every being is inclined to and tends to its proper and particular end through its proper and particular means. Inorganic nature has astronomical, physical, and chemical laws that are fixed, simple, and perfectly congruent to their end. Organic nature assumes more and more perfect laws for the sake of higher ends; and in organisms we observe organs that are most apt to their proper acts. In animals still more perfect laws thrive, and more perfect means and organs for the sake of a more excellent end; and it is even clearer in man, in whom laws, means, organs, and end are of a marvelous order.[47]

From this double harmony, dynamic and teleological, a wondrous unity results, which, as St. Thomas attests,[48] can be compared to the unity of an organism. For, in an organism, each part is for the sake of its act; for example, the eye is for the sake of seeing: so, in the parts of the universe, each creature is for the sake of its own act and its own perfection.

[47] Cf. FÉNELON, *Existence de Dieu*; MOIGNO, *Les Splendeurs de la Foi*; SAINT ELLIER, *L'ordre du monde*; FARGES, *L'idée de Dieu*, P. 1, arg. 5; SERTILLANGES, *Les sources de la croyance en Dieu*.

[48] Cf. ST. THOMAS, *Summa theologiae*, Ia, q. 65, a. 2.

In an organism, the less noble part is for the sake of the more noble; for example, the lung is for the sake of the heart; in the world, also, the less noble creatures serve the more noble; for example, those that are below man are ordered to the utility of man. Finally, in an organism, all the parts are for the sake of the perfection of the whole; thus also, each creature converges and cooperates for the perfection of the whole universe. This argument suffices for our purpose, for a fuller exposition is relegated to Theodicy.

(4) ***A Being that pre-contains all the perfections of things.*** Whatever perfection there is in the effect must preexist in the cause, either according to the same *ratio* or in a more eminent mode. But all things are effects of the Necessary *Ens*. Therefore, whatever perfection there is in things is pre-contained in the Necessary Being, either formally or in a more eminent mode. Perfections that are *simpliciter* simple—those that do not carry with them any imperfection in their concept—can be ascribed to the Necessary *Ens formally*, *as far as the thing conceived*, but not, however, *as far as the mode of conceiving*, because we conceive them as they are in things, namely, as finite and determinate, while the necessary being possesses them without limit and inexhaustibly. But mixed perfections do not belong formally to the Necessary Being, whether as far as the mode of conceiving or as far as the thing conceived, because they involve an imperfection that cannot be abstracted from their concept, but they can be attributed to the necessary being *virtually-eminently*, as to a cause of a higher and nobler order.

II. – OBJECTION: In the world there are evil things, defects, events that occur by chance. But the wise must prevent evil, defects, and events that occur by chance. *Therefore.* ***Reply.*** *I distinguish the major.* That there are evil things, defects, etc. in the world if we attend to the order of *particular* nature, I concede. That there are evil things, defects, etc. in the world if we attend to the order of *universal* nature, I deny.[49] *I*

[49] On the distinction between *particular* and *universal* nature, cf. *Summa theologiae* Ia-IIae, q. 85, a. 6: "We may speak of each corruptible thing in two ways: first, according to universal nature; second, according to particular nature. *Particular nature* is the proper active and conservative power of each thing. And according to this, every corruption and defect is against nature, as is said in Book II

distinguish the minor. That the wise man does not permit evil things which are in no way reducible to order, I concede. But that the wise man does not permit evil things that are reducible to order, I deny; for "evil things that are reducible to order," says St. Thomas, "make up the *ratio* of the beautiful." *And I deny the consequence.*

The defects that we observe in the world are evil with respect to some particular nature; thus, the fact that a certain thing ceases to be or is corrupted is evil for that thing, but with respect to the order of the universe, it is conducive to perfection.

The perfection of the universe, of course, requires that all degrees of goodness coincide in the world. But the degrees are such that some beings are indefectible and others can fall away from the good. The perfection of the universe, therefore, requires that there are defectible beings in the world. – These points are understood properly of physical evil; moral evil, however, or sin, will ultimately be punished, and thus will be reducible in some way to order, that is, to the order of justice. We shall stop here, for this is more of a theological issue.

---

of *On the Heavens*, because such a power intends the being and conservation of that to which it belongs. But *universal nature* is the active power in some universal principle of nature, such as in one of the heavenly bodies or in some superior substance, according to which even God is called 'active nature' (*natura naturans*, lit., "naturing nature") by some. This power intends the good and conservation of the universe, for which are required the alternation of generation and corruption in things. And according to this, the corruptions and defects of things are natural, not according to the inclination of form, which is the principle of being and perfection, but according to the inclination of matter, to which such form is proportionally attributed according to the distribution of the universal agent. And although every form intends perpetual being insofar as it is able, nonetheless no form of a corruptible thing can attain its own perpetuity, other than the rational soul...." – ***The Translator.***

SECOND ARTICLE

WHETHER THE WORLD IS FROM GOD BY WAY OF EMANATION

I. – WHAT EMANATIONISM IS. So far we have seen that the Author of the world is God; but one can perhaps conceive God as drawing out the world in almost the same way in which a spider spins its web. Many, in fact, thought this; this mode of origination is called "emanation." In general, "emanation" means procession of one thing from another, in whichever manner it may happen, whether out of something, or out of nothing (*ex nihilo*); emanation in the latter sense is sometimes called creation.

But emanation is defined in the stricter sense as: "The out-flowing of one substance from another." The one substance from which something comes out is called the *emanatrix*; the substance that flows out, if considered in the process of becoming, is called the *emanating*, but if it is already in existence it is called the *emanated*. This occurs in such a way that the emanated substance retains in itself something of the *emanatrix*, as the son has in himself something of the father. Emanationism, therefore, is the system that claims that the world is a certain expansion, or out-flowing, of the divine substance. Now, the emanation of the divine substance is conceived in two ways: as *immanent* and as *transient*. Immanent emanation is the evolution of God Himself, whereby God Himself expands within Himself, and becomes all things, as the pantheists claim. Immanent evolution is analogous to the evolution of a butterfly from a larva, insofar as the *same* substance, by evolving or developing itself, becomes a butterfly from a larva. Transient emanation, however, is an avulsion, a separation from the substance of God, whose terminus is the production of new substances distinct from God. Transient evolution is analogous to the production of a spider web.

II. – DEFENDERS OF EMANATIONISM. After the Egyptians and Chaldeans, Indian philosophers taught emanationism, thinking that

Brahma extracted the world out of himself, and then again absorbed it within himself; and that human souls are certain particles rent from God. The Buddhists also hold that things came out of Brahma; but they do not hold that happiness consists in the return to Brahma himself, but in the reduction of the self to nothingness, or Nirvana. The Neo-Platonists imagined that thc world emanated from God through an interminable series of intermediary beings. In other words, beings arose through invisible processions, by descending from the *One* to *Intellect, from Intellect* to *Soul*, from *Soul* to *World*. The Gnostics explained the origin of the world through a series of *Aeons*, or Demiurges, who proceeded from the Pleroma, as the book *Philosophumena* explains.

But emanationism is not dead today; it is openly taught by many philosophers and scientists. A. Sabatier says the following: "And moreover, if God has drawn creation from something that was before it, what can that antecedent thing be, if it is not God Himself? He alone has preceded all creation. Created nature is, then, the daughter of God, for it came from a seed detached from God, and God is at the same time creator and father, in the precise sense of that word. The paternity of God seems to me, then, the most rational manner, and the most in conformity with that which nature teaches us, of comprehending the origin of the world and the nature of the relations between creatures and the Creator."[50]

III. – CONCLUSION: "Emanationism is intrinsically contradictory." ***Proof.*** Emanation, as we explained, is twofold: immanent and transient. But immanent emanation leads to pantheism, whereas transient pantheism denies the simplicity of God. Emanationism, therefore, in whatever way it is proposed, is intrinsically

[50] *Philosophie de l'effort,* p. 181: "Et d'ailleurs, si Dieu a tiré la création de quelque chose qui était avant elle, quel peut bien être cet antécédent, si ce n'est Dieu lui-même? Lui seul a précédé toute création. La nature créée serait donc la fille de Dieu, puisqu'elle proviendrait d'un germe détaché de Dieu, et ce Dieu serait à la fois créateur et père, dans le sens précis que nous attachons à ce mot. La paternité de Dieu me semble donc la façon la plus rationelle et la plus conforme à ce que nous enseigne la nature de comprendre l'origine du monde et le caractère des relations, qui le rattachent au Créateur." He explains the same thing more thoroughly below, in the ninth and twelfth essays.

contradictory. ***The first part of the minor premise is established***: Pantheism is the theory that posits an evolution whereby God becomes all things. But immanent pantheism is also an expansion, or evolution, whereby God becomes all things, such that all beings are one in substance with God. Immanent emanation, therefore, is nothing but pantheism itself. ***The second part also is clear***. Transient emanation claims that something is separated and rent from the divine substance. But that which is rent necessarily consists of parts, and no longer remains simple. Transient emanation, therefore, does away with the simplicity, and consequently the infinity, of God.

IV. – OBJECTIONS.

1st Objection. That which is contained in another according to its entire *esse* emanates from the other. But creatures are contained in God according to their entire *esse*. Creatures, therefore, emanate from God. ***Reply***. *I distinguish the major*. That that which is contained in another according to its entire *esse* in a material mode emanates from the other, I concede; but that that which is contained in another in a more eminent mode emanates from the other, I deny. *I contradistinguish the minor*. That creatures are contained in God according to their entire being in a more eminent mode, I concede; but that creatures are contained in God according to their entire being in a material mode, I deny. *And I deny the consequence*. The solution is plain from what was said concerning pantheism.

2nd Objection. The *esse* that emanates is either something that belongs to the creature, or nothing at all. If nothing at all, then the creature is in no way in God; but if it is something that belongs to the creature, then, when the creature is produced, *something* is extracted from God and is received in the creature, which is the basic thesis of emanationism. ***Reply***. This *esse* that emanates does not belong to the creature as something intrinsically constitutive of it; but it can be said to belong to the creature as an extrinsic principle. – ***Reply to that which is subsumed***[51]: That a creature is in no way *in* God if the eminent *esse* is neither something that belongs to creatures nor is something intrinsically

[51] That is, reply to the minor premise of the 2nd Objection. – ***The Translator.***

constitutive of it, I concede. And, again, that a creature is in no way *from* (*de*) God if the eminent *esse* does not belong to the creature as something intrinsically constitutive of it, but belongs to the creature as an extrinsic principle, I concede. But that the creature is [thus] in no way *in* God, I deny.

A creature is in no way contained in God as a material or formal cause. Since, therefore, matter and form are intrinsic and constitutive principles, the eminent *esse* that is pre-contained in God cannot in any way be called *something that belongs to* the creature as an intrinsic element. Now, a creature pre-exists in God as in an efficient, exemplary, and final cause; but these are extrinsic principles. For a thing comes to be *by* (*ab*) an efficient cause, and not *out of* (*de*) an efficient cause. But when the efficient cause is an intrinsic constitutive principle of the thing, in this respect it is no longer called an efficient cause, but a material cause. Similarly a thing comes to be *from* (*ex*) *the imitation* of the exemplary cause, but not *out of* (*de*) an exemplary cause as a formal constitutive principle; it also comes to be *for the sake of* (*propter*) the final cause, but not *out of* (*de*) the final cause as an interior constitutive principle. Therefore, in no way can creatures be said to be *out of* (*de*) God, although they are *in* God and *by* (*ab*) God.

THIRD ARTICLE

WHETHER THE WORLD IS FROM GOD THROUGH CREATION

I. – The Notion of Creation. The term creation is used in three senses. First, to mean any origin, or production, and thus uncreated Wisdom says of itself, "I was created," that is, I had my origin, "from the beginning and before the ages."[52] Second, to mean the raising of something to a superior degree, in which sense we say that someone is made (*creari*) a doctor, a bishop, or a cardinal. Third, to mean the production of an *ens* out of nothing (*ex nihilo*). And this sense, unknown to almost all the pagans, is the genuine and proper sense. Creation, therefore, can be defined as "*the production of a thing according to its entire substance, no subject being presupposed*; or more briefly, "*the production of a thing out of nothing*." Here "*production*" is the proximate genus of the definition, in which creation coincides with generation and alteration; the rest are the proper *differentia*. The words, "*according to its entire substance*" are added in order that it may be known that whatever reality is conceived in the thing, including the matter and the form, comes to be without any subject being presupposed. In order to understand the last words, we should compare creation to the other kinds of production. Production is threefold: *creation, generation,* and *alteration*. *Generation* is the production of a substance as far as its substantial form, but not as far as its matter, as when an animal comes to be from the seed; *alteration* is the production of an accidental form, as when heat is transferred into water. Therefore, in order for creation to be distinguished from both of these, it must be the making of a substance both as far as its substantial form and as far as its prime matter.

Creation, *generation*, and *alteration* share something in common, namely, that an *ens* comes to be from *non-ens*; but they differ greatly in other respects. In alteration, an *ens* that is 'such and such' (*ens tale*) comes to be from not being 'such' (*non-ens tale*): e.g., a wall becomes white from non-white, but it does not arise from *non-ens simpliciter*; and

[52] Ecclesiasticus 24:14: "Ab initio et ante saecula creata sum."

in generation, an *ens* that is this or that (*ens hoc*) comes to be from not being this or that (*non-ens hoc*): e.g., a horse comes to be from non-horse, but it is not extracted from non-being *non-ens simpliciter*.[53] In generation and alteration, a thing comes to be out of *its* nothing, or out of not being itself, as the horse comes to be from non-horse; but not out of the nothing *of the subject*, for the horse is generated from seed that is from a subject that is presupposed; the statue comes to be from non-statue; not out of no matter at all, but out of the marble that is presupposed.[54] In a word, in generation and alteration, the thing comes to be out of *its* nothing, not out of the nothing *of the subject*; in creation, the thing comes to be *out of its nothing and out of the nothing of the subject*.

II. – It is Shown that This is the True Understanding of Creation.

Generation and alteration are the emanation of some particular *ens* from a particular cause; but creation is the making of *ens* in general by the most universal cause. But whereas something must be presupposed in the emanation of a particular *ens* from a particular cause, it is impossible to presuppose something in the making of *ens* in general from the most universal cause. Generation and alteration, therefore, are productions out of something, but creation is production that does not presuppose anything.

The minor premise [in the foregoing argument] is self-evident. If the particular cause did not presuppose anything, it would be independent in acting, and thus in *esse*, and hence it would be infinite. A particular cause, therefore, requires something previous to its action.

---

[53] *Ens tale*, a being that is such, that is, that has such and such a *quality*. Here, Hugon is contrasting the *ens tale* with the *ens hoc*, the being that has such and such a *nature*. – ***The Translator.***

[54] In generation, a thing "comes to be out of its nothing"; in other words, the absence of that thing preceded the existence of that thing. The thing does not come to be "out of the nothing of the subject"; that is, before the thing comes to be, the thing is absent, but there was nonetheless a subject (matter) present, which would later become the thing in question. – ***The Translator.***

If, on the contrary, the *esse* in general emanates, it is impossible to conceive something prior to this *esse*; and similarly, if a cause is most universal and infinite, its action must not presuppose anything at all, otherwise it would be dependent and limited in acting, and consequently in *esse*, and it would no longer retain the *ratio* of most universal cause.

Thus, creation can be said to be "*the production of a thing out of nothing, that is, nothing being presupposed*, or, *out of its nothing and out of the nothing of the subject*; or, *the production of a thing according to its entire substance*"; or: "*the production of an* ens *insofar as it is an* ens"; or, "*the transition from non-being to being* simpliciter"; or: "*the making of being in general (*esse universalissimum*) by the most universal cause*."

Now, when it is said (more briefly) to be the production out of nothing, "*nothing*" here means the negation of the material cause only, but not of the other causes. Of course, even though a created thing is not made *out of* the efficient cause, it nonetheless requires the efficient cause; even though a created thing is not extracted *out of* the exemplary cause, it nonetheless has an exemplary cause; and even though it does not arise *out of* the final cause, it is nonetheless for the sake of the final cause; but it does not have a material cause out of which it is made.

The preposition "out of" (*ex*) has two meanings:

(1) It can designate the order of succession between that which becomes and the preceding non-being (*non-esse*), as when it is said that "out of" morning comes the noon, that is, *after* the morning; (2) it excludes the material cause: "out of nothing," that is, not out of something.

From here we gather that creation is distinguished not only from generation, in which the substantial form is changed, but also from Transubstantiation, in which both the form and the matter are changed. In transubstantiation there is change, whereas in creation there is only relation (see n. VIII below); in Transubstantiation the *terminus a quo* is the substance of bread and wine, whereas in creation the *terminus a quo* is pure nothingness.[55]

[55] *Terminus a quo.* Every motion, change, or transition has two terms, or termini (plural of 'terminus'): a "terminus from which" (*terminus a quo*) it begins,

III. – Active Creation. In order that the concept of creation be fully compared to its related concepts, it is to be considered both on the part of God, or actively, and on the part of the creature, or passively. Actively, it is God's action itself.

But in God, Who is pure act, devoid of any composition, there is no distinction between acting and essence. Creation, therefore, is the essence of God itself, not taken absolutely, but *in relation to creatures.* This relation is evidently not real and predicamental, but is only one of reason. But in virtue of this relation God is said to be actively creating.

It may be objected: A relation is not of itself active. Active creation, therefore, does not denote relation. – ***Reply***: Indeed, 'relation', of itself and in its own concept,[56] is not active, but in God it is joined to some active principle, namely, to practical volition; and thus active creation can denote relation. The Angelic doctor, therefore, rightly defines active creation as "*the action of God, which is His essence, in relation to a creature.*"[57]

IV. – Is Not Creation Formally Immanent? Some authors dispute whether active creation is an immanent or a transient action. An action is said to be immanent if its terminus remains in the agent itself, but transient if the terminus lies outside of the acting principle [i.e., the agent]. In transient action three relations can be considered: first, to the agent himself, and thus it is an emergence from the agent, insofar as it denotes a respect *from*; second, to the *terminus ad quem*, and thus it is an influx of the agent insofar as it signifies a respect *to*; and third, to the *terminus a quo*, and in this way it is not an influx into the terminus, but rather a recession from the terminus, or a destruction of the terminus. Further, since in creation the terminus is not positive, there is only a recession from the terminus, that is, from nothing. The influx *from* the

---

or a starting point, and a "terminus to which" (*terminus ad quem*), or endpoint. – ***The Translator.***

[56] *In its own concept*, that is, according to, or following from, its definition. – ***The Translator.***

[57] *De potentia*, q. 3, a. 3: "Dei actionem, quae est ejus essentia, cum relatione ad creaturam."

agent has a twofold function with respect to creatures: first, the agent actualizes them, by reducing from potency to act; second, it touches the terminus, or the effect in which it is consumed. Hence, the transient action is conceived as something that exists inchoately in the agent, and consummatively in the terminus; and on account of that double formality there must be some intermediary efflux between the cause and the effect.

V. – CONCLUSION: "Creation is not a formally transient action, but only a virtually transient action."

Proof of the first part. A formally transient action is an intermediary efflux between the cause and the effect. But in creation there is no intermediary efflux between the cause and the effect. Creation, therefore, is not formally transient. The major premise is clear from what was noted above. ***Proof of the Minor.*** **(1)** ***From the texts of St. Thomas***[58]: "God acts voluntarily in the production of things, not in such a way that there is *some intermediary action of His*, as the action of the motive power in us is intermediary between the act of the will and the effect..., but it is necessary that His understanding and His willing are [identical to] His making." **(2)** ***From reason.*** An intermediary efflux is a certain motion. But in creation there is no motion. Motion, of course, is between two real termini; but in creation only the *terminus ad quem* is real; the *terminus a quo* is mere nothingness. Therefore, in creation there is no intermediate efflux. Further, this efflux cannot be in anything as in a subject: its subject cannot be God, for nothing is has pure act as its subject, but whatever is *in* God *is* God; nor can it be in the creature as in its subject, for, since this efflux is conceived as something intermediary between God and the creature, it is understood as being prior to the creature, which is its terminus. But that which is conceived to be prior to the terminus cannot be in the terminus as in its subject, for the terminus does not yet exist. Therefore, it cannot be received in the creature. ***Confirmation.*** God causes a thing through His Will and

---

[58] *Summa contra gentiles*, 4.35: "Deus agit voluntarie in rerum productione, non ita quod sit *aliqua actio ipsius media*, sicut in nobis actio virtutis motivae est media inter actum voluntatis et effectum..., sed oportet quod suum intelligere et velle sit suum facere."

Intellect. But an operation, whether of the intellect or of the will, is formally immanent. Therefore, creation is a formally immanent act.

Proof of the second part. An action is called virtually transient if, though formally immanent, it nonetheless does by its own power whatever the formally immanent action does. But creation by its own power produces whatever the formally transient action makes. *Therefore.* ***Explanation of the Minor.*** It is proper to an action that is formally transient to places a terminus outside of the agent, or to produce an external effect. But creation, by its infinite virtue, places a terminus outside of God and produces an external effect. Therefore, it is virtually transient.

VI. – Whether Creation can be Said to be Eternal. We defined creation as, "the action of God that is His essence with a relation to creatures." But creatures are produced in time, as we now suppose. Thus, we ask: Is creation eternal, or temporal?

VII. – Conclusion: "Creation on the part of God is entitatively eternal, but denominatively temporal."

Proof of the first part. Creation on the part of God is entitatively the very second act of the agent and the very act of the divine *esse*; for, in God, essence, *esse*, and operation are identified. But the second act and the divine being are eternal, otherwise in God there would be potentiality. *Therefore.*

Proof of the second part. The denomination "acting," or "creating," is in the genus of efficient cause, for it signifies a cause as being an agent in act. But the denomination "efficient cause" is not taken from the action as it is in the agent, but from the action as it is consummated in the terminus. Therefore, the denomination "creating" is taken from the divine action as it is in the terminus of creation. But the terminus of creation exists in time. Therefore, the denomination "creating" is temporal; hence, God is said to have created in time.

Explanation of the argument for the first minor premise[59]: This is the difference between formal and efficient causes: that the presence of the form in the subject suffices for the denomination "formal cause" (thus fire is said formally to be hot from the fact that the form of heat is in it), but the presence of the form in the subject does not suffice for the denomination "efficient cause," but is required that the terminus itself be attained (for fire is not said to warm something from the mere fact that heat is in the fire, but from the fact that the fire gives heat to the terminus, e.g., water). Therefore, the denomination "efficient cause" is taken from the action that is in the terminus.

VIII. – CREATION UNDERSTOOD PASSIVELY.

Creation from the part of the creature is the predicamental relation that refers the creature to God as to a principle on which it essentially depends. Of course, only two things are intermediary between action and passion, namely, change and relation: thus between the action of the fire and the passion of the water there is *change*, by reason of which water is otherwise than before, and *relation*, by virtue of which the water is related to the principle of its motion. But passive creation is not change. Therefore, it can only be a relation. *Proof of the Minor.* Change requires some real subject that is otherwise after than before. But in creation there is no subject that is otherwise after than before creation, since before creation there is nothing. Therefore, creation is not a change, but a relation.

Proof, against Scotus and Vázquez, that it is a predicamental, not transcendental, relation.

(1) ***From the authority of St. Thomas***: "That relation is an accident."[60]

(2) ***From reason.*** A transcendental relation is not distinct from the thing, but is the essence itself as ordered to another. But the creature's relation of dependence upon the Creator is not the essence itself of the creature. Therefore, that relation is not transcendental, but

[59] Namely, that the denomination "efficient cause" is not taken from the action as it is in the agent, but as it is consummated in the terminus. – ***The Translator.***

[60] *De potentia*, q. 3, a. 3, ad 3: "Illa relatio accidens est."

predicamental. ***Proof of the Minor***. The relation of dependence is founded in the creature as having existence from God. But having existence is a contingent predicate distinct from the essence, according to those things which we shall discuss concerning the distinction between essence and existence.[61] Therefore, the relation of dependence is not the created essence itself.

Thus, creation is of the genus of predicamental relation. Nonetheless, this relation is inseparable from the creature and can only be destroyed if its subject is destroyed. Of course, a relation can cease to be for three reasons: the ceasing to be of the terminus, the ceasing to be the foundation, or the ceasing to be of the subject. But the terminus of the relation here cannot cease to be, because it is God Himself; the foundation cannot cease to be, for it is the derivation of the creature from God, which derivation remains for as long as the creature lasts. Therefore, it can cease to be only by reason of the ceasing to be of the subject.

IX. – FALSE CONCEPTIONS OF CREATION. Certain spiritualist and rationalist philosophers admit, on the one hand, that the world is not produced out of preexisting matter, and that in this sense creation can be said to be out of nothing; on the other hand, they contend that the world proceeds from God as a free act comes out of a free agent, and in this respect creation cannot be said to be out of nothing. Just as we today create our free act from our substantiality, so God created the world from His power; but one difference is to be noted, that God is the supreme cause, but we are secondary causes; God created substances, whereas we only cause accidents. This is the opinion of Cousin, Franck, and others.

This notion is erroneous and provides the occasion for pantheism. We concede, of course—and we already asserted this—that creation is not from the nothing of the efficient, or exemplary, or final cause; but we have shown against the emanationists, however, that the world is not *out of* (*de*) an efficient cause, or *out of* an exemplary cause, or *out of* a final cause.

---

[61] Cf. *Ontology I.* Treatise I., q. III. art. III.

In no way, therefore, may it be conceded that the world is *out of* God.—But they add that this is altogether incongruous with our free action. Of course, intellection and volition are not creations; they are pre-contained in the subjective potency of intellect and will, and, when they are elicited, they are the actuation and modification of the faculties. Therefore, if the world proceeded from God as a free act arises from us, the world would be an actuation and modification of the divine substance, which is emanationism and pantheism. Therefore, it is to be held that creation is *out of* nothing (*ex nihilo*), even if it is not *through* nothing (*per nihilum*) or *by* nothing (*a nihilo*), or *for the sake of* nothing (*propter nihilum*).

Felix Ravaisson tried to explain creation through a certain "emptying out" of God; then resurrection follows. Just as, in the Incarnation, the Word emptied himself and, by rising again, found new life, so in creation a certain annihilation of the divine plenitude takes place; then resurrection, being, and life follow.[62] –All these things favor pantheism. For if the *esse* of creatures were to arise from the annihilation and resurrection of God, then every created reality is a certain evolution of God. And the Incarnation is not a good example, for the Word did not create anything by assuming a body, dying, or rising again.[63]

Rosmini seems to posit an indeterminate, initial *esse* that is the beginning of both God, insofar as we conceive of Him, and creatures. Further, this *esse* is somcthing that belongs to a necessary and eternal *ens*, something that belongs to the Word, which the mind of the Father does not distinguish really, but only according to reason, from the Word. The Word is the unseen matter out of which "were created all the things in the universe," as it is said in Wisdom 11:18.[64] –This doctrine is

---

[62] *Rapport sur les progrès de la philosophie en France au XIXe siècle.*

[63] VACHEROT (*Le nouveau spiritualisme*, p. 135) rightfully derides this explanation: "Dieu auteur du néant, la création expliquée par une sorte d'anéantissement suivi d'un réveil et d'une résurrection, voilá, que M. Ravaisson nous permette le mot, de ces subtilités par trop alexandrines, qui doivent rendre l'école spiritualiste indulgente pour toutes les énormités du panthéisme."

[64] DENZINGER, *Enchiridion Symbolorum*, par. 1909 [3219]. These are the old and new Denzinger numberings, respectively. – ***The Translator.***

founded on false ideological and ontological principles, which we shall elsewhere refute. But if the *esse* of creatures is something that belongs to the Word, and if the Word is the matter out of which all things are created, then we have lapsed into pantheism. For good reasons, then, this Rosminian invention was condemned by the Holy Office among 80 propositions that were proscribed on December 14, 1887.[65]

Gunther teaches that nature is a universal whole, or the universal cause and foundation that of itself makes all things and appears in each as in their phenomena, in such a way nonetheless that it is also an integral whole and that it remains as the undivided *esse* of all natural things; *and that creation is like the formation of the Non-Ego in the divine mind.* These assertions are redolent of the idealistic pantheism of Fichte; otherwise it would be unintelligible how the universal whole is an integral whole.

Finally, Gioberti did not sufficiently recede from Hegelian pantheism, insofar as he attempted to explain creation through *a certain modification of the divine idea.*

X. – Conclusion: "Not only is it not impossible for God to create something, but it is necessary to assert that God created everything."[66]

Proof of the first part, namely, that *creation is not impossible*: Since there are two kinds of impossibility, namely, intrinsic and extrinsic, creation can be considered from these two perspectives. But creation is neither intrinsically nor extrinsically impossible. Therefore, in no way is creation contradictory. ***Explanation of the Major***. Intrinsic impossibility arises from the contradiction of concepts that eliminate each other, like *circle* and *square*; but extrinsic impossibility is found in the extrinsic cause from which the thing is to be produced: thus a golden mountain is intrinsically possible, while from our perspective it cannot exist; but from God's perspective it is even extrinsically possible. ***Proof of the Minor***: "Creation" does not involve contradictory concepts, but means that a contingent effect is produced according to its entire *esse*.

---

[65] Cf. Denzinger, 1891-1930 [3201-3241].

[66] Cf. *Summa theologiae*, Ia, q. 45, a. 2.

But the fact that a contingent effect is produced by God according to its entire *esse* does not involve contradictory concepts; in fact, our mind naturally conjoins concepts of this kind; and, in contrast, it cannot understand how there can be a contingent effect that is not produced by God according to its entire *esse*. Therefore, creation is not intrinsically impossible.

Now, for it to be extrinsically possible, a cause is required that is able to produce all *esse*. But it is evident that this cause exists, for we have already shown that the maker of the world is a necessary cause and, therefore, an infinite and most universal cause. But an infinite and most universal cause must be able altogether to produce being in general, or all being (*universalissimum esse seu totum esse*).

Proof of the second part. **(1) *Indirectly*.** Either (a) the world exists through itself, whether as far as its matter alone, or as far as its matter and form; or (b) it is God Himself, or (c) distinct from God and produced by God through emanation; or, finally, it is (d) produced by God out of nothing: it is impossible to conceive another hypothesis [besides these four]. But we have abundantly proven that the world does not (a) exist through itself, whether as far as its matter or as far as its form (cf. q. 1, a. 2.). We have also shown that it is not (b) identical to God Himself (cf. a. 3.) or (c) the emanation of the divine substance (preceding Article). It remains, therefore, that (d) it has been made by God by being drawn out of nothing, or by being created.

(2) *Directly*. Everything that exists by participation is caused by that which exists by its essence. But all beings other than God exist by participation, and only God is an *ens per essentiam*. Therefore, all beings are caused by God. But God, when He causes, does not act out of something that is presupposed. Therefore, all beings are produced by God, no subject being presupposed. But to produce without presupposing anything is to create. Therefore, all things are created by God. ***Explanation of the Major***. 'A being by participation' (*ens per participationem*) means an *ens* that does not have the *ratio* of its own existence; but 'a being by its essence' (*ens per essentiam*) means an *ens* that exists essentially, in which essence and existence are one, an *ens* that is from itself (*ens a se*), a necessary being. Again, an '*ens per* *participationem*' means a finite and limited *ens*; but an '*ens per essentiam*' an infinite *ens*. Once these concepts have been understood, the truth of the major premise will be clear. An *ens per participationem* does not have in itself the *ratio* of its existence. Therefore, the *ratio* of its existence is not from another *ens per participationem*—for then we would again return to the same question—but rather from an *ens* that possesses in itself and *per se* the *ratio* of its existence, or from an *ens per essentiam*. ***Proof of the Minor.*** God is an infinite *ens* and subsistent *esse*. But an infinite *ens* cannot be but one, for who can conceive two infinite things? Similarly, subsistent *esse* must be one. Of course, *esse* is multiplied only if it is received in many subjects: hence, if it were an unreceived and unreceptive act, it would be one *esse*. Further, subsistent *esse* is an unreceived and unreceptive act, or pure act. Therefore, subsistent *esse* is one. God alone, therefore, is an *ens per essentiam*, and consequently all things other than God are *entia per participationem*. All these points are already clear from our arguments for the conclusion that the world is a contingent *ens*. All things, therefore, are produced by God. ***Proof of the Subsumed Premise***[67], namely, that God, insofar as He acts, does not extract from something presupposed: God must be independent in acting, just as in *esse*. But, if He can only act out of a presupposed subject, then he would be dependent upon something else; moreover, this subject would be prior to God's action and independent from God, and for that reason there would exist an *ens* that was not produced by God, a claim that was already refuted in the first part of the argument. Therefore, when God produces, He does not act out of something presupposed. ***Confirmation***. The mode of producing that is proper to the First Cause must be essentially distinguished from the mode of producing that is proper to secondary causes. But, if the First Cause extracts a thing only out of a presupposed subject, then it would not have a proper mode of producing; for secondary causes make substances out of a preexisting subject, as a horse generates a horse. Therefore, the mode of producing that is proper to God is to make things out of nothing presupposed.

---

[67] *The subsumed premise*, that is, the second minor premise. – ***The Translator.***

The Catholic doctrine on creation is confirmed by science. Hirn bears witness to this: "The entire universe can only be explained through the intervention of a free will, prior to every phenomenon, capable, as we have often said, not only of commanding the elements—man commands them as well to a certain degree—but capable of giving being to those elements, together with all their properties and all their qualities. *The reality of this intervention is apparent to us as a mathematical truth. Its affirmation can be seen as the last word of modern science for every upright and independent mind.*"[68]

XI. –How the Idea of Creation is Acquired. The concept of creation is handed down to us from Sacred Scripture and the Tradition of the Church. But reason alone by its own power can make evident for itself the concept of creation. For reason shows that the world does not exist through itself, that it is not God, or an emanation of God; thus, through its own argumentative power, the mind concludes that the world exists out of nothing. Further, reason proves that when God acts, He does not presuppose anything in his action. But, if He presupposes nothing whatsoever, then He takes out everything from nothing. Recall the arguments whereby the thesis was established. Given all of this, one is able to understand the definition of the Vatican Council: "If anyone should say that the One, True God, our Creator and Lord, cannot be known with certainty *by the natural light of human reason* through those things that were made, let him be anathema."[69]

---

[68] Hirn, *La Vie future et la Science*: "Tout l'ensemble de l'univers ne peut s'expliquer que par l'intervention d'une volonté libre, antérieure à tout phénomène, capable, non pas seulement, comme on l'a dit si souvent, de commander aux éléments—l'homme leur commande aussi dans une certaine mesure—mais capable de donner l'être à ces éléments avec toutes leurs propiétés et toutes leurs qualités. *La réalité de cette intervention nous apparaît comme une vérité mathématique. Son affirmation peut être regardée comme le dernier mot de la science moderne pour tout esprit droit et indépendant.*"

[69] *De revelatione*, can. 1: "Si quis dixerit Deum unum et verum, *Creatorem* et Dominum nostrum, per ea quae facta sunt, *naturali rationis humanae lumine* certo cognosci non posse; anathema sit." Denzinger 1806 (3026). – ***The Translator.***

In actual fact, however, there were very few philosophers who, lacking the aid of revelation, attained the genuine notion of creation. Lucretius says: "the principle from which we shall begin: Nothing is ever divinely created out of nothing."[70] But perhaps he received that notion from tradition.

Aristotle's principles on *act* and on the *prime Mover* logically contain the doctrine of creation, although it is not clear from his thinking. The texts in which he claims that the world is unproduced and eternal create a difficulty,[71] but these are interpreted in the sense that the world, although created by God, exists from eternity, and did not have a generation or beginning.

XII. – Difficulties Resolved.

1st Objection. It is a common dictum among philosophers that, "nothing comes to be out of nothing" (*ex nihilo nihil fit*). But creation is out of nothing. Therefore, creation does not occur, or is impossible. ***Reply***. *I distinguish the major*. That nothing comes to be out of the nothing [i.e., the absence] of any cause, whether efficient, material, exemplary, or final, I concede; and that nothing is made by particular causes out of the nothing [i.e., the absence] of a material cause, I concede; but that nothing is made by the most universal cause out of nothing, I deny. *I contradistinguish the minor*. That creation occurs out of the nothing of a material cause, I concede; but that creation occurs out of the nothing of any cause, I deny. *And I deny the consequence.*

In the realm of secondary causes, becoming is a certain motion, insofar as things emanate from particular causes through a transmutation from one thing into another. But every transmutation requires a subject that is otherwise after than before. Therefore, nothing is made by particular causes without presupposing a subject. But, as we have already proven, it is necessary that the most universal causes do not presuppose a subject.

---

[70] *De rerum natura*, Book 1, vv. 148-149: "principium hinc cujus nobis exordia sumet: Nullam rem e nihilo gigni divinitus unquam."

[71] Cf. *De coelo et mundo*, 1.10. Cf. St. Thomas, *In I De coelo*, lect. 22 (Text 102). – ***The Translator.***

2nd ***Objection***. Between nothingness and an *ens* there is an infinite distance. But an infinite distance cannot be covered. Therefore, it is impossible for an *ens* to be taken out of nothing. ***Reply***. *I distinguish the major.* That there is a *negatively* infinite distance between nothingness and an *ens*, I concede; but that there is a *positively* infinite distance between these, I deny. *I contradistinguish the minor.* That a *positively* infinite distance cannot be covered, I concede; but that a *negatively* infinite distance cannot be covered, I deny. *And I deny the consequence.*

Distance is said to be "*positively* infinite" when there are infinite intermediaries between two extremes, and, evidently, this distance cannot be covered. But, in creation, there is no infinite intermediary between nothingness and *ens*. The distance, therefore, is not positively infinite, but only *negatively* so, that is, from *non ens simpliciter* to *ens simpliciter*. Further, just as the transition from not being *this* (*non ens hoc*) to being *this* (*ens hoc*) can occur through a particular cause, so the transition from not *non ens simpliciter* to *ens simpliciter* can occur through the most universal cause.

3rd ***Objection***. In every production, becoming (*fieri*) is prior to having been made (*factum esse*). But where becoming is prior to having been made, there must be a subject, and thus a creation that is said to occur without a subject is contradictory. ***Reply***. *I distinguish the major.* That, in every *successive* production, becoming is prior to having been made, I concede; but that this is the case in every *instantaneous* production, I deny. *I concede the minor premise. I distinguish the conclusion.* That in every successive production there must be some subject, I concede; but that this is the case in every instantaneous action, I deny. *And I deny the last conclusion*, for creation is an instantaneous production.

A successive production is first in the process of being done (*in fieri*) before having been done (*in facto esse*); thus, the process of being heated is prior to having been heated. But an instantaneous action simultaneously is being done and has been done; for example, to be in the process of being understood and to have been understood occur simultaneously. Therefore, since creation is not a successive change, in it being done and having been done are simultaneous.

4th Objection. If the world comes to be out of nothing, then either there is an intermediary instant between *esse* and *non esse*, or there is not. If there is not, then the world both will be and will not be in the same instant, which is absurd; if there is some intermediary instant, another absurdity follows, because in the same instant the world neither is nor is not. Therefore, the production of the world out of nothing is contradictory. ***Reply***. Between *esse* and *non esse* there is no real intermediary instant, but an imaginary instant can be conceived. For an instant is an accident of a durable, successive *ens*. But an accident is not prior to its subject. Therefore, a real instant prior to being cannot be conceived; but the first instant begins with the reality itself of the created *ens*. Just as before creation an imaginary time can be conceived, so an imaginary instant can be thought of at the end of which *non ens* can be posited; and thus suffices that the world be conceived as *esse* and *non esse* in the same instant. Further, just as imaginary time does not share continuity with real time, so the imaginary instant must not be thought of as sharing continuity with the real instant through some medium. This objection, therefore, proceeds from a false imagination, because we conceive of creation in the manner of a successive action in which many instances concur. But creation is an instantaneous production in which coming to be and having become are simultaneous.

FOURTH ARTICLE

ON THE SUBJECT AND THE OBJECT OF CREATION

I. – QUESTION. It is certainly *de fide* that no creature creates or has created. But certain philosophers, especially among the Gnostics and Manichees, have imagined that the First Cause did not create immediately, but by means of inferior intelligences, which for that reason are called demiurges (*demiourgoi*).

The Albigensians renewed this error, contending with the Manichees that the visible world was created by a certain evil principle.

Avicenna also claimed that, even though creating is proper to the first cause, an inferior cause also can create in virtue of the first cause: for example, "a separated substance creates another after itself and the substance of the world and its soul, and [then] the substance of the world creates the matter of inferior bodies."[72] Catholics almost unanimously repudiate the idea that a creature can create as a principal cause; but many think that it is possible, given God's absolute power, for there to exist a noble creature to which principally belongs the role of creating inferior beings with divine assistance. This is the opinion of Durandus, which Arriaga thinks is probable. But Peter Lombard, Durandus, Arriaga, Hurtado, Vásquez, and Suárez hold that a creature can create in the manner of an instrument, given God's absolute power. The Scholastics deny it most unanimously.

---

[72] *Summa theologiae* Ia, q. 45, a. 5: "quod ... substantia separata... creat aliam post se, et substantiam orbis, et animam eius; et quod substantia orbis creat materiam inferiorum corporum."

II. – First Conclusion: "The idea that a creature can create as a principal cause is contradictory."

1st Argument. The order of causes corresponds to the order of effects. Therefore, the most universal effect is proper only to the most universal cause. But *esse simpliciter* is the most universal effect. Therefore, the production of *esse simpliciter* is ascribed only to the most universal cause. But to produce *esse simpliciter* is to create. Therefore, creating is proper only to the most universal cause, which is God alone.

Esse simpliciter is rightly called the most universal effect, for in that case it is not this or that *ens* that is produced, or even an *ens* with respect to one of its parts only, as when a plant arises out of a seed, but *esse* absolutely, *esse* in its entirety, or all the realities that are conceived in it. But *ens* absolutely, in its entirety, is manifestly the most universal effect, in which nothing at all is presupposed. Hence it is that, in every creaturely operation in which *esse* is produced, God must supply his assistance so that *esse* may be bestowed. Therefore, just as the creature cannot act or conserve its *esse* independently from the divine assistance, so it cannot produce *esse simpliciter* by creating.

2nd Argument. Creation requires infinite power. But infinite power cannot be given to creatures. *Therefore.* ***Proof of the Major.*** If we consider the entire series of beings (or creatable being in all its latitude), which is infinite, it is manifest that an infinite power is required for creation; but the Angelic Doctor shows that infinite power is required even for the creation of a single creature.[73] The power of the maker is not considered only from the point of view of the substance of the thing made, but also from the point of view of the mode of making, for a greater heat does not only heat *more*, but also heats *more quickly*. Therefore, although creating some finite effect does not entail an infinite power, nonetheless creating it out of nothing does entail an infinite power. For the more potency is removed from act, the greater the power required in the agent: the power engaged in burning a green tree is greater because in it the potency to burn is more remote from act than in a dry tree.

---

[73] Cf. *Summa theologiae* Ia, q. 45, a. 5, ad 3.

And so, an infinite power is required wherever the potency is infinitely removed from act.

But in creation, potency is infinitely removed from act, since there is no potency, nor proportion of *non ens* to *ens*. Therefore, creation requires an infinite power in its agent.

3rd ***Argument***. The principal cause of a thing must have, as its proper effect, everything in which is found the formal *ratio* of that thing which it causes. This is clear, for if anything that has the formal *ratio* of that thing is not made by the cause, then already this thing is not the first and principal cause, but, at most, a secondary and instrumental cause. Therefore, the principal cause of *esse* has as its proper effect everything in which is found the formal *ratio* of caused *esse*. But the formal *ratio* of caused *esse* is found even in the creating creature. Therefore, if a creature were the principal cause of being, it would have itself as its proper effect, against the axiom: "Nothing is the cause of itself" (*Nihil est causa sui*).[74]

III. – Second Conclusion: "It is contradictory for a creature to create even as an instrumental cause."

Proof. A secondary instrumental cause participates in the action of a superior cause only insofar as it acts dispositively towards the production of the effect of the principal agent, through something proper to itself. But in creation no creature can act dispositively for the sake of the effect of the principal agent. Therefore, no creature can be an instrument of creation. ***Explanation of the Major***. Unless the instrument previously has something dispositive towards the effect of the principal cause, whether on the part of the thing made or on the part of its mode of making it, it would be employed in vain, and there would not necessarily be diverse instruments for different actions. It is altogether required, therefore, that the instrument have its proper effect, which is prior, at least in nature, to the effect of the principal cause, and that it be a sort of preparation for it, as it shall be explained in the treatise on causes. ***Proof of the Minor***. The action of the instrument is an accident, since every created action belongs to the genus of accident;

[74] Cf. *Summa contra gentiles* 2.21.

for this reason, it requires a subject that is prior to it at least in nature in order to receive and act dispositively. But creation does not admit a preexisting subject. Therefore, no creature can act dispositively towards the effect of creation.

In other words, when the effect of the principal cause is presupposed in the effect of the secondary cause, it is clear that the secondary cause does not act dispositively toward the effect of the first cause. But the proper effect of God the Creator is presupposed in any effect whatsoever of any secondary cause. Therefore, no secondary cause can act dispositively toward the effect of God the Creator. ***The minor premise is clear***. The effect of God the Creator is most universal *esse*. But *esse* is presupposed in all things and nothing is presupposed in it; before it there is nothing that can be created through the act of an instrumental cause.[75]

IV. – OBJECTION: The action of an instrument presupposes a subject *in which it inheres*, but not necessarily a subject *out of which it comes*. But creation is said to be only without a subject *from which* it comes out. Therefore, from the fact that the action of the instrument is in the subject *in which it inheres*, it does not follow that it cannot act dispositively toward creation. ***Reply***: From the very fact that an action is an accident, it follows not only that it must be dependent on the subject in which it inheres, but it also requires a subject on the part of the thing that is done. For insofar as it is an accident, it has a subject. But the action that has a subject does not make the whole, but requires something that precedes it, and on which it depends and out of which it comes. Therefore, the action of the instrument, even on the part of the thing made, presupposes a subject in which it inheres and on which it depends and out of which it comes.

Of course, an accident is no more independent in acting than in *esse*. But it depends on a subject in *esse*. Therefore, it depends on a subject in acting. But to depend on a subject in acting is to come out of

[75] Because creation does not have a subject *from which* the created thing is drawn out. – ***The Translator.***

a subject. Therefore, an accident can only act if it comes out of a subject.

V. – NOTE. It is to be noted that a creature can be understood to be an instrument of creation in two senses. First, by its very nature it is the instrument of the divine power for creating, as heat is by its nature fire's instrument for generating heat. Second, as an instrument, not through a natural proportion, but from the divine efficacy which raises the creature, in the way that the Sacraments are instruments that produce grace.

Now, it is certain that no creature by its very nature can be the instrument of creation; otherwise it would naturally have an efficacious instrumental power with respect to all *esse*. And no theologian can be cited that asserts such a thing. The dispute, therefore, is only concerning instruments taken in the second sense. Also, an instrument understood in this sense must do something previously, otherwise the thing employed could be called the occasion, or merely the means, but not the instrumental cause that is concurring in one and the same operation along with the principal cause. Further, the action of the principal agent and of the instrument are not two distinct operations, but one, as the writer and the pen have principally one effect. Therefore, since even the [supernaturally] elevated creature cannot concur through something previous in the same operation with God the Creator, it is contradictory for a creature, even if elevated, to be the instrument of God the creator.

Further, it is necessary to note three things here:

(1) The proper and natural operation of an instrument implies two things. First, that the proper operation of the principal agent may be communicated to the instrument and become its *ratio*. Second, that the instrument may properly and truly apply the power of the principal agent to the effect, because, if the instrument does not apply anything of itself, then it is a mere occasion in the presence of which the principal agent does all things immediately and of itself.

(2) The proper and anteriorly dispositive operation of the instrument cannot be supplied by the absolute power of God. For the preceding operation is in the genus of material and dispositive cause.

But God cannot supply a material cause; He certainly can make, in the manner of an efficient cause, something that supplies the material cause, but he cannot take on the role of a material and dispositive cause; otherwise imperfection would be ascribed to God. Therefore, it is impossible for a dispositive operation to be supplied by God. Therefore, neither can a creature become an instrument of creation through the absolute power of God.

(3) A creature that in no way acts cannot be assumed as God's instrument; e.g., God cannot assume a tree in order to resurrect the dead, as a tree does not have any operation connatural to itself. God certainly can restore a dead man to life in the presence of a tree that does not do anything, but the tree cannot be the instrument of resurrection unless it acts by its own power dispositively (e.g., by touching) towards the action of God, Who is the One resurrecting the dead man.[76]

VI. – WHICH THINGS ARE PROPERLY CREATED. We treat here of the object of creation.

Two conditions are required for something to be created in the proper sense: one on the part of the terminus from which (*terminus a quo*), and the other on the part of the terminus to which (*terminus ad quem*). On the one hand, it is necessary that no aspect whatsoever of the things that are created be presupposed, but that whatever they have has come out of nothing; on the other hand, it is necessary that the same things receive, by virtue of their production, not only *esse secundum quid*, or partial *esse*, but their entire *esse*, namely, whatever there is in them that pertains to the *ratio* of an *ens*.

VII. – [THIRD] CONCLUSION: "Only composed and subsistent things are properly created."

Since becoming is the way to *esse*, becoming belongs to those things to which *esse* belongs. But to be created is a certain kind of becoming. Therefore, to be created belongs properly to those things to which *esse* belongs. But *esse* belongs properly to subsistent and composite things. Therefore, so does being created. ***The major premise***

[76] Cf. BÁÑEZ, *In Iam Partem*, q. 45, a. 5.

is an axiom: For becoming is ordered to *esse* as to a terminus and an end. But that which is ordered to an end is specified by it. Therefore, becoming is also specified by *esse*. Hence the principle: "Each thing has as much becoming as it has being" (*Unumquodque habet fieri sicut et esse*). But being created, although it is not becoming in the strict sense because it does not involve motion, nonetheless, according to our manner of conceiving it, it is understood as a certain kind of becoming. Hence, to be created belongs to those things to which it belongs to be (*esse*). ***But we say in the subsumed premise***: "*Esse* belongs properly to composite and subsistent things." For form and accidents and other things of the sort are not that *which* has *esse*, but only that *by means of which* (*quo*) something is. That which properly has *esse* is a composite and subsistent thing. Hence, form and accidents, as they are rather *coexistents* than beings; thus, they are said to be rather *co-created* than created; but that which is properly created is a subsistent thing.[77]

VIII. – Five Corollaries. From these principles the following points result.

(1) Grace and supernatural gifts are not created, for they are accidents; they, therefore, *inhere* (*insunt*) rather than are (*sunt*): they do not come to be *simpliciter* out of nothing, but come to be in their dependence upon the subject in which they are sustained. If a subject is created with its accidents, as the soul of Christ was created with its grace and the Gifts, those beings can be said to have been co-created, but not in the proper sense. For that is properly said to be co-created which is created together with the subject, as a passion that is proper to the subject. But the gifts are not proper passions of the subject. Therefore, they are not co-created, properly speaking, even though they are made simultaneously. But if they are founded on an already-existing subject, as it occurs in us, they are in no way co-created, but are drawn out of the *obediential* potency of the subject. The theologians discuss this issue.

(2) Prime matter is not created, properly speaking and *per se*, but is co-created as dependent upon the form in its *esse*. For matter does not receive *esse*, properly speaking and *per se*, but only insofar as it depends

[77] Cf. *Summa theologiae*, Ia, q. 45, a. 4.

upon the form. But each thing has as much becoming as it has being. Therefore, matter is not created or made, properly speaking and *per se*, but only insofar as it depends upon form.

(3) In the first production of things, form is not created, properly speaking (i.e., drawn out from the potency of matter), but is co-created towards the production of the whole. It is not created. Form, of course, is not that *which* is (*id quod est*), but only that *by which* something is (*id quo aliquid est*). – It is not drawn out. The leading out of forms from matter is done through a transmutation of matter. But the transmutation of matter presupposes form; for matter exists and is transmuted only insofar as it is connected to form. Therefore, the first forms are not drawn out of matter. It remains, then, that they are co-created towards the production of the whole.

(4) All subsistent things (i.e., things that exist without a subject) are created, properly speaking; e.g., the angel and the human soul. The very fact that they exclude a subject upon which they depend and from which they are drawn out necessitates that they be made out of nothing; and, because they have *esse*, properly speaking, and subsist in themselves, they have becoming, properly speaking, and so they are created, properly speaking. Even though the human soul is united to the body, it nonetheless is independent from the body in its being, and consequently in its becoming; for this reason it is made and created *per se*.

(5) Subsistent composites, even as composites, are created, properly speaking, in the manner of a whole that is made primarily and *per se*. For, since creation is the production of *esse simpliciter*, or of the entire *esse* [of a thing], it follows that that is primarily and *per se* created which primarily and *per se* is a whole and has *esse* in the manner of a whole. But that which is a whole and has *esse* in the manner of a whole is a composite as composite, and not a composite in the sense of a part, because the part in the composite does not exist *per se*. Therefore, that which is created, properly speaking, is the composite, not in the sense of a part, but as composite, or in the manner of a whole. ***Confirmation***. Creation, as a most perfect production, must not proceed from the imperfect to the perfect, but must immediately attain the perfect. But a production that proceeds from the parts to the whole proceeds from the imperfect to the perfect; for indeed, the whole is greater and more perfect than the parts. Therefore, creation does not begin with the parts and then composes the whole, but immediately attains the whole, that is, the composite as composite, or as having the *ratio* of the whole. Now, the parts are made by the same action by which the whole is produced. Hence, the whole results from the parts, not as being presupposed in becoming, but only as composing and constituting its *esse* in fact. Therefore, it is not the case that the parts are made beforehand and then out of them the whole is produced, but rather the whole, and the parts within the whole, are made immediately. This is the opinion of the Thomists, [in particular, that of] Cajetan, Báñez, and John of St. Thomas.

FIFTH ARTICLE

WHETHER CREATION IS A FREE ACT OF GOD

I. – ERRORS. Among the rationalists that believe in creation, many deny God's freedom in creating, and assert that God could not have created. This is the opinion of Emilius Saisset, Cousin, and Robinet. Other philosophers, even if unconscious, lapsed into the same error, thinking that God cannot choose among possible worlds the one he chose, but was necessarily forced to choose the most perfect. This is the opinion of Leibnitz, Wolff, etc. Others, such as Bouillier and Jules Simon, contend that the world can progress infinitely and become absolutely perfect until it cannot become any more perfect.

II. – FIRST CONCLUSION: "Creation is altogether free."

Proof, from St. Thomas.[78] The necessity of creation would arise either from an intrinsic cause, that is, intrinsic to the nature of the world, or from an extrinsic cause. But neither of these can be affirmed. *Therefore.*

Proof of the Minor [in Parts].

(1) It cannot be from the intrinsic nature of the world. For it does not belong to the world to exist of itself, but by virtue of another, as we have abundantly proven. But it is impossible for that which does not draw *esse* from itself to have, of itself, necessity of *esse*. *Therefore.*

(2) An extrinsic cause is either efficient or final; but necessary creation comes neither from the efficient cause nor from the final cause. The necessity of creation, therefore, is not derived from any extrinsic cause.

Proof of the [Subsumed] Minor[79] ***[in Parts].***

---

[78] Cf. *Summa contra gentiles*, 2. 31.

[79] The subsumed minor is the minor of the argument labeled as (2), namely, that "necessary creation comes neither (A) from the efficient cause nor (B) from the final cause." Since this premise has two parts, Hugon will prove each part separately (A and B). Note that the proof for part (A) is a complex sorites involving multiple levels of subsumed premises. – ***The Translator.***

(A) A necessary effect follows from the efficient cause when the agent acts out of a necessity of nature. But God does not act out of a necessity of nature. *Therefore.* – It is clear that God does not act out of a necessity of nature. For we have shown that the author of the world is endowed with intellect and will. But every agent enjoys freedom through intellect and will. *Therefore.* ***Further [Proof of the Subsumed Minor]***: the first and most perfect act must be ascribed to God, the first agent. But a free act concerning external objects is prior and more perfect than a necessary action. From this it is clear that those things amongst us which act voluntarily are more perfect than those which act through a necessity of nature. Therefore, to God must be ascribed a free action concerning objects external to Himself, such as the world. – ***Explanation of the [Subsumed] Minor***. We said: *concerning external objects*, a free operation is more perfect than a necessary one. With respect to one's own and intrinsic good, a free action of God is not better than a necessary action. For, since the intrinsic divine good is infinite, like God himself, it is desired necessarily; and it is not better to remain indifferent to it; indeed, if God could be indifferent with respect to His own good, it would follow that the divine good would be limited and participated. But with respect to an extraneous good, a free action is better and greater than a necessary action. For he who elicits a necessary action concerning an external object is moved by the object, and so he is inferior to the object; for indeed, nothing is determined except by something stronger and superior; but he who is indifferent and free with respect to the object, by that very fact is considered greater and superior to the object. But it is more perfect to be *simpliciter* superior to an object than to be inferior to the object, as is evident. Therefore, he who elicits a free action concerning an object, or concerning an extraneous good, exists more perfectly *simpliciter* than he who has a necessary and unavoidable action. It is clear, therefore, that creation is not necessary on the part of the efficient cause.

(B) The same is to be said concerning the final cause. Things are necessary on the part of the end when the end either altogether cannot be attained without those things, e.g., the conservation of life without food, or cannot be well attained without those things, e.g., a journey without a horse. But the end of the divine will can be both attained *simpliciter* and attained well without creation. *Therefore.* For the end that the divine will intends in the production of things is not something outside of God, but is God Himself, or the divine goodness. But the divine goodness is not originated by creatures, whether with respect to its being, because it is *per se* and necessary, or with respect to its well being, because it is perfect *simpliciter* and according to itself as well as fully self-sufficient. Therefore, the end of the divine will can be both attained *simpliciter* and also well-attained without creatures.

III. – Difficulties Resolved.

1st Objection. Indifference belongs to the essence of freedom. But in God no indifference can be conceived; otherwise God would not be pure act. Therefore, in God there is no freedom. ***Reply.*** *I distinguish the major.* That *active* indifference belongs to the essence of freedom, I concede; but that *passive* indifference belongs to the essence of freedom, I deny. *I contradictinguish the minor.* [That in God no *passive* indifference can be conceived, I concede; but that in God no *active* indifference can be conceived, I deny.] *And I deny the consequence.* As we shall discuss in Psychology, since passive indifference is a capacity to receive many determinations, it designates a potentiality and an imperfection; therefore, it does not belong to the essence of liberty, and cannot be ascribed to God in any sense. But active indifference is a power or perfection that is not conceived as indeterminate or in potency, but which can produce indifferently diverse effects, and which retains such eminence to diverse things, that it cannot be forced to one. This belongs to the essence of freedom and it is found in God. Even though the divine will cannot be conceived to be indeterminate and in potency to diverse determinations, it has, nonetheless such eminence to diverse things that its effect could indifferently be the existence or the non-existence of creatures. Since the world exhibits neither an infinite dignity by which it must necessarily be desired, nor a total defect by which it must necessarily be despised, God is objectively indifferent to willing or not willing the existence of the world.

2nd ***Objection***. God wills His own goodness together with that of creatures. But He is not indifferent to willing His own goodness. Therefore, nor is he indifferent to willing the goodness of creatures. ***Reply***. *I distinguish the major.* That He wills His goodness together with that of creatures, but in a diverse manner, I concede; but that He does so out of the same motive, I deny. *I concede the minor, and I deny the consequence.* Certainly, the act by which God wills His own goodness is the same as that whereby He wills creatures, but of couse the *ratio* of willing it is different. For the goodness of God is the proper and adequate objcct of the divine will. He wills, therefore, His own goodness necessarily, just as every faculty has a necessary relationship to its proper object. But He wills other things insofar as they are ordered to His goodness as to an end. Now, we do not will that which is a means to an end unless without them the end cannot be attained. But the end of the divine will, as was said, can be both attained *simpliciter* and also well-attained without creatures. Therefore, God does not will creatures necessarily.

3rd ***Objection***. The creative action of God is eternal. But to an eternal action corresponds an eternal effect. Therefore, the effect of creation is eternal and necessary. ***Reply***. *I distinguish the major.* That His creative action is eternal entitatively and according to that which it implies in the agent, I concede. But that it is so denominatively and insofar as it implies something concerning the effect, I deny. *I contradistinguish the minor.* That to an action that is eternal in both ways—both according to what it implies in the agent and as it implies something concerning the effect—there corresponds an eternal effect, I concede. But that an eternal effect corresponds to an eternal action only in the first manner, I deny. *And I deny the consequence.* An action is considered in two ways. First, according to that which it implies in the agent, that is, insofar as it is the second act of the agent; for this reason creation is necessary and eternal, because it is a divine act and the Divine *Esse* itself. Second, insofar as it implies something concerning the effect; furthermore, the effect follows necessarily once the action has been posited, if the action is formally transient, because then the action essentially implies an outward (*ad extra*) efflux. But if the action is formally immanent and only virtually transient, the effect does not follow necessarily from the positing of the action, but only insofar as it determines the agent himself. Now, creation is formally immanent and virtually transient. "Therefore, even though the action of the first agent is eternal, it is not necessary that His effect be eternal," says the Angelic Doctor.[80]

4th Objection. If the effect of an action is only temporal, a certain change follows. But the idea that there is a change in God is contradictory. Therefore, the effect of creation must be eternal. ***Reply***. *I distinguish the major.* That a certain *extrinsic* change follows, I concede; but that an *intrinsic* change follows, I deny. *I contradistinguish the minor.* [That an *intrinsic* change in God is contradictory, I concede; but that an *extrinsic* change in God is contradictory, I deny.] *And I deny the consequence.* When the world is produced in time, God does not change any more than a tree changes when it is known by the mind. Before, the tree was hidden from me, and now I know the tree; the whole change is on the part of my intellect; the tree is not affected intrinsically, but only extrinsically, insofar as my cognition, which did not exist before, now corresponds to it. Thus, in creation all change is on the part of the creature; the action is affected only extrinsically, insofar as to it corresponds some effect in time which did not correspond to it from eternity.

5th Objection. The highest good (*summum bonum*) must communicate itself. But God is the *summum bonum* from all eternity. Therefore, He communicates Himself from eternity. ***Reply***. *I distinguish the major.* That the *summum bonum* must communicate itself, either necessarily or freely, I concede. *I subdistinguish the term "necessarily."* That he communicates himself necessarily inwardly (*ad intra*), I concede; but that He must do so necessarily *ad extra*, I deny. *I concede the minor premise. I distinguish the consequence*. That God communicates Himself eternally *ad intra*, I concede; but that he does so *ad extra*, I deny. The *summum bonum* communicates Himself in the highest manner, necessarily and eternally to Himself *ad intra*. This

---

[80] *Summa contra gentiles* 2.35: "Nec tamen oportet quod, si primi agentis actio sit aeterna, quod eius effectus sit aeternus."

communication constitutes the divine processions *ad intra* of which Theology studies. But when one says that, "the highest good is diffusive of Himself" (*Summum bonum est diffusivum sui*), even *ad extra*, the sense is not that He has an actual need to communicate Himself, but only that He can, when He wills. Of course, He would not be the *summum bonum* unless He enjoyed the freedom whereby He could diffuse His perfection according to His will.

IV. – SECOND CONCLUSION: "The present world in some true sense can be said to be the best and most perfect world."

First, ours is the best world on the part of the end, insofar as it cannot be ordered to anything better than to God and His extrinsic glory. Second, ours is the best world for attaining that degree of glory which God wished to obtain through the creation of our world. This is the doctrine of St. Thomas:

> "These things being supposed, the universe cannot be better on account of the most fitting order given to these things by God, in Whom the good of the universe consists. If any of these things were better, the proportion of the order would be corrupted, just as if one string were to be struck more than is due, the melody of the guitar would be corrupted."[81]

Proof from Reason. God is an agent of infinite wisdom and infinite virtue. But when a wise agent intends some end absolutely, he must, if he can, choose the means that are most conducive to attaining that end. Therefore, God must have chosen the most apt means for attaining the degree of glory that He intended. Therefore, if there could be a more perfect world, the aforesaid degree of glory would be lost, and the proportion of order would be eliminated, just as a melody is corrupted when a string is struck more than is due.

From these things one can better understand the axiom, "It belongs to the best to make the best things" (*Optimi est optima facere*), that is,

[81] *Summa theologiae* Ia, q. 25, a. 6, ad 3: "[U]niversum, suppositis istis rebus, non potest esse melius; propter decentissimum ordinem his rebus attributum a Deo, in quo bonum universi consistit. Quorum si unum aliquod esset melius, corrumperetur proportio ordinis, sicut, si una chorda plus debito intenderetur, corrumperetur citharae melodia."

the best things in the order of their intended end, although it does not belong necessarily to the best to produce the best things absolutely. Hence follows the Third Conclusion:

V. – Third Conclusion: "God could have, and still can, create some other world more perfect than ours."[82]

***Proof.* 1st *Argument*.** No *ens* that is finite as far as its essence can participate in God in an infinite way. But the present world, no matter how perfect, is finite as far as its essence. Therefore, our world participates in God only in a finite way. But God can be participated infinitely. Therefore, between our world and the infinite participability of God there is a positively infinite distance. Furthermore, between things that are infinitely distant from each other there can be an infinite number of intermediaries. Therefore, between God and our world there can be an infinite number of intermediary worlds that participate in God's perfection more and more, and there will never be a world so perfect that it cannot further participate in the divine perfection.

For this reason, even if a continuous gradation [of perfection] is admitted in creatures, there cannot be a creature that attains the most perfect degree of all. Rather, each creature has its own degree of [essential] perfection, such that each creature cannot exceed its degree without that creature being destroyed [and another, higher creature generated].

***2nd Argument*.** An infinite power cannot be exhausted by a finite effect. But the power of God is infinite, and the present world is a finite effect. Therefore, the potency of God is not exhausted by the present world, and hence he is able to make more and more worlds without end.

***3rd Argument*.** Optimism discards divine freedom. For, if God must necessarily make that which is best, then He must necessarily create, supposing that it is better to create than not to create; He must necessarily conserve the world, supposing that it is better to conserve it than to annihilate it, etc., etc.

---

[82] Cf. FÉNELON, *Réfutation du système du P. Malebranche*, ch. 6; BONIFAS, *Etude sur la Théodicée de Leibnitz*.

VI. – On the Oneness of the World. It has been shown that an infinite number of worlds are possible; but now we ask whether many worlds exist or can exist that do not have any unity among themselves.

Solution. (1) There are not, nor can there be, many worlds that have no unity of order or of end among themselves. For, if many things have a relationship to the same efficient, exemplary, and final cause, then they will exhibit some unity of order and of end among themselves. But all created things and all creatable things necessarily have a relationship to the same efficient, exemplary, and final cause, namely, God. Therefore, there are not, nor can there be, many worlds that do not coincide in a unity of order and of end. Further, all created and creatable things must fall in some degree of *ens*, whether in *esse*, or in living, or in sensing, or in understanding. Therefore, it is necessary that all created and creatable things come together in some unity. This is held unanimously.

(2) It is not congruent for there to be many worlds that do not exercise some operation among themselves.

It is necessary that all worlds coincide in the same end. But the tendency towards an end arises through operation. Therefore, it is necessary that all worlds coincide in some operation, insofar as some act upon others, just as it is in fact evident that all planetary systems exert some attraction among themselves.

The *ratio* of the minor premise is that a creature is not its own end, by virtue of its nature. Therefore, it can only tend to its end and attain its end by acting.

VII. – From this we Deduce the Following: If God had created only two bodies, they would have to touch each other immediately. As it is evident from what has been said, they would coincide in some operation and would act upon each other. But bodies can transmit their operation to each other only through contact: either through immediate contact or through the contact of a third, intermediary body. Therefore, the two aforesaid bodies would have to touch each other. But it could not be through the contact of a third, intermediary body, because we are supposing that there are no bodies other than the two. Therefore, it must be through immediate contact.

VIII. – Whether there are Many Inhabited Worlds, or Many Planets in which Inhabitants Live.

Solution. In itself it is not contradictory, but there is no reason to believe that this is the case. Science has not observed any certain vestiges of life in other planets; indeed, it admits that in many planets the climate is not fit for the conditions of life, although in certain planets, like Mars, life is possible. The arguments from its fittingness are not cogent; for the glory of God can be sufficiently procured through angels, human souls, and especially through the Incarnation of the Son of God among men.

If there were rational creatures in other planets, would they not be ordered to eternal life just as we are? There is no reason for supposing that they were left in the state of pure nature, for theologians do not admit that such a state has ever occurred in fact, even though they do think it is possible.

But it seems altogether fitting that God would reveal to us something about the existence of these beings, who would have to be our fellow citizens in eternal life. Therefore, just as He made manifest to us the existence of the angels, because they are co-participants with us of the same beatitude, so, *it would seem*, He would have revealed the existence of rational creatures living on other planets. This argument, even though reasonable, is not absolutely convincing, however, for God is not obliged to give the aforesaid revelation, or to elevate those creatures to the supernatural order.

IX. – The Doctrine of the Church Regarding Creation. We shall resolve the question of the origin of the world through a brief and clear synopsis of the Catholic doctrine, and we shall remit the rest to the theologians.

The Church teaches that:

(1) God is the *one principle* of all things, distinct from the world.

(2) He created *out of nothing* (*ex nihilo*) first angelic and mundane creatures, then human creatures.

(3) He did so, not out of necessity, but with free will.

(4) Not out of *need*, to increase His Beatitude or to acquire it, but to *manifest* his perfection through the goods which are bestowed on creatures.

(5) *Not from eternity*, but *from the beginning of time*.

(6) He made angels and the world simultaneously. This last point is a doctrine that is altogether certain, although it does not seem to have been defined expressly and directly.

Here is the text of the Vatican Council, which confirms and expands the definition of the Fourth Lateran Council:

> "This sole true God by His goodness and 'omnipotent power', not to gain or increase His own beatitude, but to manifest His perfection by the good things which He bestows on creatures, with a most free will, 'simultaneously from the beginning of time made out of nothing both spiritual and corporeal creatures, namely angelic and mundane, and then the human creation, common as it were, constituted out of both spirit and body'."[83]
>
> Canon I: "If someone should deny the One, True God, the Creator and Lord of all things visible and invisible, let him be anathema."[84]
>
> Canon V: "If someone does not confess that the world and all the things that are contained in it, both spiritual and material, were produced by God from nothing as regards to their whole substance, or shall have said that God created not by a will free of all necessity, but that he created as necessarily as He necessarily loves Himself, or shall have

---

[83] DENZINGER 1783 (3002): "Hic solus verus Deus bonitate sua et 'omnipotenti virtute' non ad augendam suam beatitudinem nec ad acquirendam, sed ad manifestandam perfectionem suam per bona, quae creaturis impertitur, liberrimo consilio 'simul ab initio temporis utramque de nihilo condidit creaturam, spiritualem et corporalem, angelicam videlicet et mundanam, ac deinde humanam quasi communem ex spiritu et corpore constitutam' (Conc. Lateran. IV: cf Denzinger 800 [1530]). – ***The Translator.***

[84] DENZINGER 1801 (3021): Si quis unum verum Deum visibilium et invisibilium creatorem et Dominum negaverit: anathema sit. – ***The Translator.***

denied that the world was created for the glory of God: let him be anathema."[85]

Canons III and IV attack and condemn pantheism in all its forms.[86]

[85] IDEM 1805 (3025): Si quis non confiteatur, mundum resque omnes, quae in eo continentur, et spirituales et materiales secundum totam suam substantiam a Deo ex nihilo esse productas, aut Deum dixerit non voluntate ab omni necessitate libera, sed tam necessario creasse, quam necessario amat se ipsum, aut mundum ad Dei gloriam conditum esse negaverit: an. s. – ***The Translator.***

[86] Cf. IDEM 1803-4 (3023-24). – ***The Translator.*** On creation, see S. THOMAS, Ia, qq. 45ff; *Summa contra gentiles* 2; *De potentia*; S. BONAVENTURE, in II. *Sentent.*; CAPREOLUS, *In II Sent.*; CAJETANUS, BÁÑEZ, *In Iam Partem*; JOHN OF ST. THOMAS, *Cursus theologicus* (edit. VIVES), t. 2 and t. 4; SUÁREZ, *Disput. metaphys.*; FARGES, *L'Idée de Dieu*; VACANT, *Etudes théologiques sur les Constitutions du C. du Vatican*; DUILHÉ DE SAINT-PROJET, *Apologie scientifique...*; MAZZELLA, *De Deo creante*; JANSSENS, *Summ. Theol.*, tom. 2.

THIRD QUESTION

On the Duration of the World

Having determined what is the origin of the world, now it is fitting for us to ask questions regarding the duration of the world: Can it exist from eternity? – How old it is? – Will it last forever?

FIRST ARTICLE

WHETHER THE WORLD COULD HAVE EXISTED FROM ETERNITY

I. – OPINIONS. We have shown that creation from eternity is not necessary, whether it be considered on the part of God or on the part of creatures. Indeed, we know by faith, that is, from the Book of Genesis and from the declarations of the Church in the Fourth Lateran and Vatican Councils, that the world was created in time: "*From the beginning of time* He created creatures out of nothing...."

The traditions of different peoples are consonant with this faith.

Certain philosophers thought that the beginning and newness of the world could be demonstrated with certainty from the characteristics of the world itself. Geology, of course, makes evident that man did not always exist on earth, and that animals and plants had a beginning, for it is clear that there was a lifeless (*azoica*) period prior to the age of living things. But science cannot prove with certainty whether the matter of the world, or the primeval clowdy mass, had or lacked a beginning. For this reason, we hold by faith and tradition that the world, taken as a whole, includes the primitive mass, had a beginning.[87]

---

[87] Cf. S. THOMAS, Ia, q. 46, a. 2; P. SERTILLANGES, *Revue Thomiste* 5, p. 746.

At present, therefore, we treat only of the possibility of eternal creation. The Fathers certainly denied it, but in the sense that the pagans and heretics defended it. These heretics claimed that creatures are eternal almost by their own right and with a natural eternity; the Fathers replied that no creature was eternal; in fact, they claimed that it was impossible for them to be eternal, because not even an omnipotent God can bring about an eternal and uncreated creature.

The Scholastics, however, treat the question in a different sense, that of gratuitous, participated and dependent eternity.

Boethius, St. Thomas, Durandus, Cajetan, Báñez, Suárez, Wolff and many Neo-Scholastics, such as Sanseverino and Liberatore, defend the possibility of eternal creation understood in this sense. Some admit this possibility only for permanent things, but not for successive beings. This is the opinion of Durandus, John of St. Thomas, Goudin and others.

But St. Anselm, Hervaeus Natalis, Henry of Ghent, Toledo and many modern authors hold the negative view. It is also attributed to St. Albert, but perhaps undeservingly. St. Bonaventure is also cited as being in favor of it, but the Seraphic Doctor sometimes seems have an inclination towards the opposite view.[88]

II. – CONCLUSION: "It cannot be demonstrated apodictically that the concept of an eternal creation is contradictory."

The terms of the conclusion should be understood correctly. We do not say, "the possibility of eternal creation is clearly demonstrated," but only, "the impossibility is not proven evidently and apodictically"; in other words, the reasons that are brought forth against the possibility of an eternal creation, "are not necessarily conclusive, although they are probable," as the Angelic Doctor says.[89]

Moreover, we do not claim that the being of each permanent or successive creature is possibly [eternal], because perhaps the particular *ratio* of human creature is an exception, but claim that in general the

---

[88] Cf. BONAVENTURE, *In II Sent.* d. 2, a. 1, q. 6.

[89] ST. THOMAS, *Summa contra gentiles* 2.38: "non... de necessitate concludunt, licet probabilitatem habeant...."

world could have been created from eternity, as far as certain beings, whether permanent or successive.

Proof of the Conclusion. 1st Argument. The principle of demonstration is *that which is* (*quod quid est*), or the essence of the thing according to the *ratio* of its species. But each thing according to the *ratio* of its species abstracts from time; for this reason, the essences of things are said to be always and everywhere. Therefore, the principle of demonstration abstracts from all time, and hence it cannot be demonstrated that some creature cannot exist always and from eternity. – ***Objection***: The essence of a thing certainly abstracts from this or that time, but not from all time in general. – ***Reply***. Time in general is an accident of a created thing: for a creature does not require by virtue of its essence that it last and remain in *esse*. But an accident is not included in the concept of the essence. Therefore, time in general is not included in the concept of the essence of a created thing. Therefore, the essence does not abstract from this or that time, but from time in general altogether.

2nd Argument. Impossibility is twofold, namely, extrinsic and intrinsic. But neither is incompatible with eternal creation. *Therefore.* Extrinsic impossibility comes from a defect in the power of the agent. But God was not of lesser power in eternity than now: it would be absurd for someone to think that God grows in power with time. Therefore, an eternal creature is extrinsically possible. Now, intrinsic possibility is not an accidental predicate of a thing, but is included in its very essential principles. But essential principles belong to a thing independently of time. Therefore, intrinsic possibility belongs to a thing independently of time. Therefore, an eternal creature is intrinsically possible. ***Explanation of the Minor***. Essential principles belong to a thing independently from accidents. But time is an accident. Therefore, essential principles belong to a thing independently of time.

III. – An Objection is Precluded.

Objection: An eternal creature is intrinsically possible in the sense that it is intrinsically possible for a creature to exist in some determination of time, but not in the sense that it is intrinsically possible for it to exist eternally. – ***Reply***. If a creature were considered intrinsically possible only in the sense that it can exist in time, then it would not be called possible absolutely but only possible in time: and thus intrinsic possibility is measured by time. But intrinsic possibility is included among the essential principles of a thing. Therefore, the essential principles of a thing are measured by time; and because time is an accidental predicate, it follows that essential principles are measured by accidental principles, which is absurd.

Hence, the argument is again established: Intrinsic possibility is included in the essence of the thing. But the essence does not involve time (which is an accident) in its concept. Therefore, intrinsic possibility does not include time in its concept. Therefore, a thing is not to be called intrinsically possible merely because it can exist in time, but absolutely possible, abstracting from any time whatsoever.

IV. – ADVERSARIES ARGUE BY WAY OF *REDUCTIO AD ABSURDUM*. Adversaries, then, cannot prove the contradiction of an eternal creation with a positive demonstration that uses as a middle term the quiddity of a thing. By the very fact that they posit time, they add something that is beyond the quiddity of a thing. They try, therefore, to have recourse to a *reductio ad absurdum* argument. But, on the contrary, St. Thomas showed the non-contradiction [of eternal creation] by an argument that uses the notion of quiddity itself and of essential principles in this way: Quiddity and essential principles abstract from time altogether. Therefore, the possibility of a created thing abstracts from any time whatsoever.

Now, an argument that proceeds by way of *reductio ad absurdum* cannot prevail against one that is founded on the very notion of quiddity. Therefore, even if the *reductio* arguments cannot be resolved, the mind ought not for that reason dissent from the conclusion that is derived from the notion of quiddity; because then we would have a reason to suspect that some sort of sophistry is hidden within these *reductio* arguments.

V. – CONCLUSION: "An absurdity does not follow from the opinion of St. Thomas." An absurdity would arise if the action did not precede the effect; or if the creature were equated to God; or if the *ratio* of creation, namely, the transition from non-being to being, were not preserved; or if the action of God were not free; or if there were an infinite number [of actual things]; or if one could add to the infinite. But in none of these respects is there an absurdity [in St. Thomas]. *Therefore*. ***The major*** stands on sufficient enumeration. ***The minor*** is proved in parts:

(1) In a successive action, the cause must certainly precede its effect *in duration*; but that does not belong to the *ratio* of cause as such. For when the cause acts instantaneously, as the operative power acts together with its cause in the same instant, then the operation and the effect can take place in the same instant, and consequently it is sufficient that the cause precede the effect with a priority of nature. But creation is an instantaneous action. Therefore, it is sufficient that the cause precede its effect with a priority of nature; and this would certainly be preserved even if creation were from eternity.

(2) Creatures would not for this reason be equated to God in eternity, "because the divine *esse* exists entirely all at once without succession, but this is not the case with the world."[90] – On this hypothesis a creature would only lack *act* in the beginning, but not *aptitude*. Hence this duration would be an accidental eternity and not an essential eternity such as that which belongs to God, because it would not imply the necessity and indefectibility of *esse* in virtue of the creature's essence.

(3) Creation in its essential concept signifies only that things are not made out of a presupposed subject. But even if creation were eternal, things would not be made out of a presupposed subject. Therefore, the *ratio* of creation would be preserved. ***Objection***: Creation requires a transition from non-being (*non esse*) to *esse*. Therefore, the *esse* of the creature must be under 'non-being' (*sub non-esse*) before being under 'being' (*sub esse*). Further, the prepositional phrase "out of nothing" (*ex nihilo*) implies an order of succession between nothing and *esse*. Therefore, the *ratio* of creation is preserved only if nothingness

[90] *Summa theologiae* Ia, q. 46, a. 2, ad 5: "[Q]uia esse divinum est esse totum simul, absque successione; non autem sic est de mundo."

precedes *esse*. ***Reply***: It belongs to the *ratio* of a creature that it *can* begin, that it *can* transition from non-being to being; but not that it *actually* begins, that it *actually* transitions from non-being (*non esse*) to *esse*; on this account it is sufficiently distinguished from God, who *cannot* begin or be conceived as transitioning from non-being (*non esse*) to *esse*. ***Further***, in order for the creature to transition from non-being (*non-esse*) to *esse*, a priority of *ratio* or of nature is sufficient; and the succession between *esse* and non-being (*non-esse*) that the preposition "out of" (*ex*) implies is one of *ratio* and of nature only. Moreover, this priority would exist even if things existed eternally; for the creature would be understood as being under 'non-being' (*sub non-esse*) prior to being under 'being' (*sub esse*).

(4) The free decision of God is not made in time, but from eternity, since He does not need to take any time or counsel for deliberating. Hence, creation would be necessary only on the supposition arising from the divine will, and this necessity does not preclude freedom.

(5) The greatest difficulty on which adversaries especially insist is that eternal creation seems to introduce an actually infinite multitude. One can reply that there is no such problem in the case of creatures in which there is no motion or succession; but the objection arises again concerning beings that undergo succession and the reply is not obvious.

1st Objection. If lions have come to be throughout all eternity, then there is an actually infinite number of lions; if men have come to be throughout all eternity, then there is an actually infinite number of human souls, etc., etc. There shall also be an infinite number of motions, and the stars shall have orbited an infinite number of times, etc. But such multiplicity, being a number that is actually infinite, is impossible. *Therefore*. ***Reply***. It must be noted that two senses of eternity can be distinguished: eternity of *permanence* and eternity of *succession*. Eternity of *permanence* consists in each thing being determined from eternity. But eternity of *succession* does not imply that this particular motion, or this particular orbital cycle of the Sun, or that the coming to be of this particular lion is from eternity (this is indeed impossible, because a particular motion, and the coming to be of a particular animal require something prior in duration, namely, something mobile, something that generates the animal, and dispositions); rather, it only means that before any motion, or before any coming to be, there must be another motion, or another coming to be. Hence, it cannot be said that all motions and all coming to be, whether all taken collectively or each taken singly, are from eternity. In other words, a posterior part, or even a particular prior part, could not be eternal; but before any given part (*pars assignata vel assignabile*) there would be more and more parts, and we would never be able to arrive at the first. The matter shall become clearer if we compare it to eternity *a parte post*.[91] The thoughts of the soul and of angels will be eternal *a parte post*, not because any one of them will be eternal, but because after any one in particular there can follow another, and yet another, without ever coming to a last one. Many difficulties that are frequently cited, e.g., that the years, weeks, days, cannot be eternal, vanish thanks to this explanation. Of course, any day in particular is not eternal, but all taken collectively are eternal, insofar as before any determinate day there is always another, and another, and before any determinate hour there is another, and before a determinate moment there is another, etc. Hence it is clear how an actually infinite multitude is avoided. For it would only be a successive infinity. But successive infinity consists of finite elements, because it is made up of parts, one of which is added to the other. Therefore, it contains only finite elements. But when a finite thing is added to another, the result is not something actually infinite. Therefore, it would not be an actually infinite multitude. The Angelic Doctor says the following on the matter: "Something infinite, even if it is not all at once in act, can nonetheless be in succession, since any infinite thing understood thus is something [actually] finite."[92]

---

[91] *A parte post*, or *ex parte post*, that is, eternity in the *future*, or the lack of an *ending* in time. This notion is contrasted to that of eternity *a parte ante*, or *ex parte ante*, eternity in the *past*, or the lack of a *beginning* in time. Thus we could say that the human soul is not eternal *a parte ante*, for it created in a given moment in time, but it is eternal *a parte post*, because it has no temporal ending, due to its immortality. – ***The Translator.***

[92] *Summa contra gentiles* 2.38: "[I]nfinitum, etsi non sit simul in actu, potest tamen esse in successione: quia sic quodlibet infinitum acceptum finitum est."

2nd Objection: The infinity of space is something actually infinite. Therefore, the infinity of time implies something actually infinite. ***Reply***: This objection is not difficult to handle. Space is a permanent quantity all of whose parts exist simultaneously; hence, if space were infinite, it would be something actually infinite. But time is a successive quantity whose parts do not coexist simultaneously. Therefore, even if time were not limited by a terminus *a parte ante*, nonetheless it would always have a terminus *a parte post*. – If the generations of lions were eternal, lions would not be for that reason infinite in number, because they would cease to be successively. This objection is more difficult concerning human souls, if indeed souls do not cease to be. Absolutely speaking, God can annihilate existing souls as He produces new souls. This is certainly not fitting, but nonetheless it would not be contradictory; and thus the impossible scenario of an actually infinite multitude would be avoided. It can also be added that the *per accidens* coming to be of man from eternity is contradictory, due to a special difficulty that does not apply to the rest of things. "One must consider the fact, however, that this is a particular argument. Hence, someone could say that the world is eternal, or at least that some creatures are eternal, such as the angels, but not man."[93] From this text, many authors, even among Thomists, infer that the Angelic doctor does not admit the possibility of an eternal creation of successive things. But they are mistaken. In this reply St. Thomas only means that the objection does not apply to angels; in no way does he deny the possibility of an eternal creation of successive things. In the sixth objection he says this: "If the world always was, then an infinite number of days preceded this one. But the infinite cannot occur. Therefore, this day would never have happened, which is manifestly false."[94] The objection evidently concerns successive beings. Therefore, [if these authors were right] he should have replied: "I deny that assumption,

---

[93] *Summa theologiae* Ia, q. 46, a. 2, ad 8: "Considerandum tamen quod haec ratio particularis est. Unde posset dicere aliquis quod mundus fuit aeternus, vel saltem aliqua creatura, ut Angelus; non autem homo."

[94] *Ibid.*, arg. 6: "Si mundus semper fuit, infiniti dies praecesserunt diem istum. Sed infinita non est pertransire. Ergo nunquam fuisset perventum ad hunc diem, quod est manifeste falsum."

because our opinion regards permanent beings." But instead he admits the assumption and replies: "In reply to the sixth argument it must be said that a change is always understood as being from one terminus to another. Now, however many days may have passed between one day and another, there is going to be only a finite number of days [between them]. But the objection proceeds as if, given the extremes, there could be an infinite number of intermediaries."[95] In the *Summa contra gentiles*, he speaks also of successive beings: "It would follow that one can add to the infinite if to [an already infinite number of] past days or rotations were added daily more and more [days or rotations]."[96] He replies: "Nothing prevents there being an addition to something insofar as it is finite. Supposing that time is eternal, it would follow that it is infinite *ex parte ante*, but finite *ex parte post*, for the present is the terminus of the past."[97] Therefore, the Angelic Doctor defends the possibility of eternal creation for successive beings in which *days*, *orbits*, etc. are distinguished. He concedes the impossibility with respect to man, not under the *ratio* of successive being as such, but under the *ratio* of an *incorruptible successive being*. Nonetheless, many authors, such as Capreolus, affirm that even the human species could have existed from eternity.[98] But these philosophers seem either to appeal to miracles or to posit an actual infinity. And even if these miracles are not contradictory, they would nonetheless occur despite an intrinsic impossibility. Others reply that the souls that are successively created never make up an actually infinite multitude, because that which is added to a preexisting number is always a finite part: but a finite thing added to another does not make up an infinite thing. There would therefore never be anything infinite *a parte post*.

---

[95] *Summa theologiae* Ia, q. 46, a. 2, ad 6: "Ad sextum dicendum quod transitus semper intelligitur a termino in terminum. Quaecumque autem praeterita dies signetur, ab illa usque ad istam sunt finiti dies, qui pertransiri poterunt. Obiectio autem procedit ac si, positis extremis, sint media infinita."

[96] *Summa contra gentiles* 2.38: "Sequitur quod infinito fiat additio: cum ad dies vel circulationes praeteritas quotidie de novo addatur."

[97] *Ibid.* "[N]ihil prohibet infinito ex ea parte additionem fieri qua est finitum. Ex hoc autem quod ponitur tempus aeternum, sequitur quod sit infinitum ex parte ante, sed finitum ex parte post: nam praesens est terminus praeteriti."

[98] *In II Sent.* d. 1, q. 1 (ed. PABAN/PEGUES, t. 3, p. 25).

(6) An absurdity does not arise from the fact that something is added to the infinite, as we just heard from the Angelic Doctor. One cannot add anything to the infinite *simpliciter*, but one can add to that which is infinite *secundum quid*, insofar as it is finite.

Now, it has been shown that in our hypothesis there is nothing infinite *simpliciter*, but only something infinite *secundum quid*, or rather something indefinite.

VI. – WE CONCLUDE, therefore, that the arguments of our adversaries are only probable. Many modern authors speak all too confidently by presenting the negative opinion as *certain*, as if they had discovered new arguments that make the matter evident. But no new arguments are offered; rather, our adversaries repeat the very same objections that the Angelic Doctor raised against himself.

We give an end to this issue by appropriating the words of St. Thomas: If it is evident that eternal creation is contradictory, then "it is astonishing how Augustine and the noblest of philosophers [including the Angelic Doctor himself, the prince of theologians and the norm for philosophers] did not see this contradiction."[99]

[99] *De aeternitate mundi* (*Opusculum* 27): "Mirum est etiam quomodo nobilissimi philosophorum hanc repugnantiam non viderunt." Cf. Fr. SERTILLANGES, *Revue Thomiste*, September and November 1897, January 1898; and among the older authors, CAPREOLUS, *loc. cit.* CAJETAN, BAÑEZ, *In Iam Partem*, q. 46; GUÉRINOIS, *Clypeus Phil. Thomist.*, MAILHAT, *Phys.*

SECOND ARTICLE

HOW OLD IS THE UNIVERSE

OR

ON COSMOGONY AND GEOGONY

I. – Laplace's Hypothesis. Since the world in fact did not exist from eternity, we must consider how many years have passed since it was first formed. An *a priori* argument cannot be established, and similarly science cannot establish any certain conclusions. But the hypothesis that scientists have unanimously embraced, following Laplace, is most probable.[100] It can be presented thus: in the beginning there was a certain immense nebulous or gaseous mass, like an impalpable fluid, or an extremely tenuous dust of imperceptible rarefaction. Into this mass were introduced a vehement impulse, and a double, simultaneous motion: namely, that of rotation and that of translation. Now, from the motion, heat and light were generated. But as each substance acted with such fast impetus, matter was condensed toward the center; and, because centrifugal force increases with motion, it gradually came about that an immense ring was torn away from the mass. Here again by the motion of the impulse a new sphere came to be, from which, on account of the same laws, a ring was torn away. This event was repeated many times: thus were formed many spheres, and each of their satellites, which transitioned from a gaseous state into a solid state.

As heat decreased, many of these spheres obtained solid surfaces and thus passed from being suns to being planets. Our Earth, therefore,

[100] Hugon only means this was the most probable scientific hypothesis at the time of writing (1903 for the 1st ed.; 1927 for the 3rd ed.). As progress is made throughout history in the field of empirical science, its hypotheses tend to become more or less probable. The philosophical principles that Hugon has laid out so far, however, are in no way dependent on Laplace's hypothesis. – ***The Translator.***

underwent three phases: a gaseous phase, or the state of being a nebulous mass, then the phase of being a sun, and then the phase of being a planet.

II. – THIS COSMOGONY, OR MODE OF EXPLAINING THE FORMATION OF THE WORLD, CANNOT AT ALL BE PROVEN FALSE BY PHILOSOPHERS. If it is admitted that God created the first mass, and that he gave it the impulse and the power whereby it could evolve, this view does not have any contradictions; on the contrary, God's power shines forth all the more, in that it can make so many and diverse globes out of a single mass; and also His wisdom and goodness insofar as he does not wish to make all things alone, but gives to creatures the dignity of being causes, and employs them to bring about wonderful things. This evolution requires almost innumerable days and years, but in all of this there is no contradiction. On the contrary, "it seems fitting," says Pesch, "that God left as a certain vestige of eternity these extremely long periods of time, as he left the wondrous extensions of the heavens as a vestige of his immensity."[101]

III. – THIS OPINION IS ALSO SCIENTIFICALLY CONVINCING. "A charming laboratory experiment done by Plateau, a wise Belgian physicist, allows us to reproduce in a small scale, but in a striking way, the grand phenomena that we have just discussed. One mixes water and alcohol in a jar in such a proportion that, when onc adds droplets of oil in the mix, they have the same density as the mixture and remain in suspension in the form of a sphere. One can then regard this sphere as not being subject to the force of gravity. One then stirs the water around this sphere with an iron rod, to which a clock-like motion imparts a motion of rotation that gradually accelerates. One then sees that the small sphere flattens at its poles and inflates at its equator, and then that oily rings successively detach from it: these reunite in small spheres that

[101] PESCH, S.J., *Phil. nat.* Lib. II, disp. 1, sect. 2: "[C]ongruum videtur reliquisse Deum quasi vestigium quoddam aeternitatis suae longissima temporum spatia, sicut reliquit vestigium immensitatis miras coelorum extensiones."

continue to turn upon themselves and around the central mass in the manner of satellites."[102]

IV. – Geogony. We must add a few concerning geogony, that is, the manner in which our Earth was formed. We saw how the Earth obtained a solid surface, which in the beginning was tenuous and was disrupted more than once by convulsions: but gradually, as the globe cooled, it became solid. (The crust of the Earth probably did not exceed 30 km [or about 19 miles] in depth; rather, many have established that it was only 20 km [or about 12 miles].) But while the planet cooled, gaseous vapors turned into water; hence the seas. And the Earth gradually attained its present state due the fact that external cold and internal heat interacted, such that continents and mountains successively formed, etc.

V. – Geological Ages. Thus, the Earth underwent many phases before perfect animals and man himself appeared. These phases can be ordered according to the geological strata in the following manner:

(1) The *Primitive* or *Azoic* age, which left no certain vestiges of life.
(2) The *Primary* or *Paleozoic* age, in which living things are found, but which are very dissimilar to the animals of our time. The principal formations of this time are:

---

[102] Maisonneuve, *Géologie.* Introduction: "Une charmante expérience de laboratoire due à un savant physicien belge, Plateau, nous permet de reproduire en petit, mais d'une façon saisissante, les phénomènes grandioses qui viennent d'être exposés. On mélange, dans un bocal, de l'eau et de l'alcool, en proportion telle que des gouttes d'huile, versées avec précaution dans ce liquide, aient la même densité que lui et y restent en suspension sous forme d'une sphère. Celle-ci peut donc être regardée comme soustraite à l'action de la pesanteur. On traverse cette sphère avec une tige de fer, à laquelle un mécanisme d'horlogerie imprime un mouvement de rotation de plus en plus accéleré. On voit alors la petite sphère s'aplatir à ses pôles et se renfler à l'équateur, puis des anneaux huileux s'en détacher successivement: ceux-ci se réunissent en petites sphères qui continuent à tourner sur elles-mêmes et autour de la masse centrale comme autant de satellites."

(a) The *Cambrian* formation, where the principal vegetable organisms are algae; and the principal animals are worms, trilobites, and crustaceans;

(b) The *Silurian* formation, where the vegetable organisms are the fucoid algae and the animals are the trilobites, the mollusks, and certain fishes;

(c) The *Devonian* formation, whose vegetable organisms are the ferns, and the animals are the polyps and fishes;

(d) The *Carboniferous* formation, whose vegetable organisms were the ferns and the coniferous plants; and the animals are the spiders, the scorpions, the labyrinthodonts, etc.;

(e) The *Permian* formation, in which terrestrial reptiles appear.

(3) The *Secondary* or *Mesozoic* age, in which new types of living things appear. The principal formations are:

(a) The *Triassic* formation, whose vegetable organisms are the ferns, the coniferous plants, large horsetails; the animals are echinoderms, brachiopods, oysters, labyrinthodonts, dinosaurs, and marsupials.

(b) The *Jurassic* formation, whose principal vegetable organisms are cycads, yew trees, and monocotyledons; and the animals are the marsupials, the ammonites, the belemnites, dinosaurs, atlantosaurs, apatosaurs, ichthyosaurs, plesiosaurs; the first birds, the archaeopteryx and the pterodactyl.

(c) And the *Cretaceous* formation, whose vegetable organisms are the cycads, the conifers, poplar trees, beech trees; and the animals are the brachiopods, the mollusks, the cephalopods, the crustaceans, the crocodiles, the iguanodons. Among the enyaliosaurs, the most eminent was the mosasaur; also birds endowed with teeth are found, such as the *hesperornis* and the *ichthyornis*.

(4) The *Tertiary* or *Neozoic* age, in which living things, reptiles, fishes, mammals, are very similar to the living things of our time. The principal formations are:

(a) The *Eocene* formation, whose vegetable organisms are laurels, palm trees, acacias; and the animals are the large mammals, the *palaeotherium*, the *xiphodon*; among fishes, the rays.

(b) The *Oligocene* formation, whose vegetable organisms are the laurels, maple trees, the cinnamon; and the animals are the *palaeotherium*, hippopotamuses, echinoderms, etc.

(c) The *Miocene* formation, whose vegetable organisms are poplar trees, sycamore trees, the cinnamon, grassy plants; and the animals are the ruminants, the mastodon, the rhinoceros, simians, the beaver, the *hipparion*.

(d) The *Pliocene* formation, in which the vegetable organisms are nearly the same as in our age; but the animals are the *hipparion*, the rhinoceros, the bear, the wolf, the cow, the deer, the *deinotherium*.

(5) The *Quaternary* age, in which the vestige of man is found. At the beginning of this period there was an *ice age*, after which deluges and floods followed. Among the species of animals that were extant then, some still remain today in the same places, others have migrated to warmer regions; others have become altogether extinct, such as the woolly mammoth , the woolly rhinoceros, the *hippopotamus major*, the Irish Elk, the *mylodon*, the glyptodont, the giant moa, etc.

VI. – How Many Years the Formation of the Earth Lasted. This is impossible to estimate; in fact, it is uncertain how much time elapsed from the beginning of life in the primary age until now. De Lapparent thinks that the duration can be reasonably estimated to be between twenty million and a hundred million years. Dana speculates that it is forty eight million years: that is, thirty six million years for the primary age; nine million years for the secondary age; and three million years for the tertiary age. This calculation is especially good for America. Not all scientists, however, admit that the duration of the primary age exceeds that of the other ages by as much as Dana supposes. Concerning man, nothing proves with certainty that his antiquity exceeds ten thousand years.

Let it be noted, however, that all these things that we have said concerning cosmogony and geogony do not have a demonstrative, but a problematic, value.[103]

Now, Sacred Scripture leaves the question intact.[104] The biblical chronology is not resolved in all its aspects, and the Hebrew text does not agree with the Septuagint version. Further, if we admit the reckoning of the Septuagint, the antiquity of man is confined to eight or ten thousand years,[105] and this space of time is completely sufficient for aptly explaining both the historical testimonies and the scientific facts that can be brought to the issue. "Let it be necessary to adopt the chronology of the Septuagint, since it gives us a notably longer time, we are convinced of it, but we do not perceive any reason to extend that chronology beyond the eight or ten thousand years that it gives us as a maximum."[106]

---

[103] That is, a probable or dialectical value. – ***The Translator.***

[104] That is, if one assumes that the six "days" of creation of Genesis 1 represent ages, and not natural (24-hour) days, which is the view of some Fathers of the Church, such as St. Augustine. – ***The Translator.***

[105] See the genealogies in Genesis 4, 5, and 11. – ***The Translator.***

[106] HAMARD, *La Science et l'Apologie chrétienne*, p. 31: "Qu'il soit nécessaire d'adopter la chronologie des Septante, puisqu'elle nous fournit un temps notablement plus long, nous en sommes convaincus, mais nous n'apercevons aucune raison d'étendre cette chronologie au-delà des huit ou dix mille ans qu'elle nous accorde comme un maximum."

THIRD ARTICLE

WHETHER THE WORLD IS TO LAST FOREVER

I. – First Conclusion: "The world will never be reduced to nothing."

God can certainly annihilate the world. For just as before things were He was able not to communicate *esse* to them, and thus not make them, so after they are already made he is able not to put being into them, and thus they would cease to be, which is the same as reducing them to nothing.[107]

From revelation, however, we know that in fact nothing will be annihilated by God.

Proof. What is done by God to a creature either happens to it according to the natural course of things, or miraculously beyond the inner order of things. But annihilation occurs neither according to the natural order nor miraculously. Therefore, in no way does annihilation occur. ***Proof of the First Part of the Minor***. That which God is going to do according to the inner, natural order of things can be considered from the perspective of the natures themselves. But from the natures of creatures we can demonstrate that that none of them is reduced to nothing, either because they are immaterial and thus in them there is no potency for non-being, or because they are material and thus remain at least as matter, which is incorruptible, insofar as it is the existing subject of generation and corruption. Therefore, according to the natural order God is not going to annihilate creatures. ***Proof of the Second Part of the Minor***. That which is done miraculously is ordered to the manifestation of grace, for the natural order must only be changed for the sake of a higher order, which is the order of grace. But reducing something to nothing does not belong to the manifestation of grace, since the divine power and goodness is shown by the fact that things are conserved in *esse*. Therefore, God does not reduce things to nothing by

[107] Cf. *Summa theologiae* Ia, q. 104, art. 3.

miracle. Hence, it is to be said *simpliciter* that He will not reduce anything at all to nothing.[108]

II. – Second Conclusion: "The corporeal world will not always remain in the state in which it is now."

Against Hesiod, Plato, and many others among the ancients, we say "the corporeal world," for the spiritual world is altogether incorruptible; nor do we speak of Heaven, but of the world insofar as it includes our Earth and the inferior heavens [or heavenly bodies, i.e., planets, starts, etc.], which, according to spectral analysis are made out of corruptible elements like our planet.

Proof of the Conclusion. Just as our Earth, which first underwent a solar phase, and gradually after the heat was exhausted obtained a solid surface, so it seems that the Sun and the stars after many ages gradually will obtain a solid surface, after their heat is exhausted. But after the heat and light of the Sun and stars is exhausted there will remain no possibility of life, whether on the Earth or on any of the other planets. Therefore, at some point the world will cease to exist in the present state. "Thus, in the beginning, a nebula that condenses; in the end, a series of dark worlds placed in such conditions that none of the material phenomena that we know can take place, death replaces life, darkness light: such is the picture that modern science authorizes us to conceive."[109]

III. – How Will the Ruin of the World Happen?

The teaching of Scripture and Catholic Tradition bear witness that there will be a universal conflagration: "But the day of the Lord shall come as a thief, in which the heavens shall pass away with great violence, and the elements shall be melted with heat, and the earth and

---

[108] Cf. *Ibid.* a. 4.

[109] De Lapparent, *Discours sure les enseignements philosophiques de la science*: "Ainsi, à l'origine, une nébuleuse qui se condense; à la fin une série de globes obscurs placés dans des conditions telles qu'aucun des phénomènes matériels que nous connaissons ne peut s'y accomplir, la mort substituée à la vie, l'obscurité à la lumière: tel est le tableau que la science moderne nous autorise à concevoir."

the works which are in it shall be burnt up."[110] And the Church sings: "Free me, Lord, from eternal death, that fearful day when the heavens and the Earth will be changed, when Thou shall come to judge the world through fire."[111]

The same tradition is found among the pagans, both poets and philosophers. Among the poets, Ovid in particular is cited: "He remembered that by fate there shall also be a time in which the sea and the land, and the royal heaven, snatched up, shall burn, and the mass of the world, besieged, will be in distress."[112] As is Lucan: "If these peoples, Cesar, do not burn by fire, they shall burn with the Earth; they shall burn with the whirlpool of Pontus; the whole world is left in a funeral pyre."[113] Among the philosophers, Cicero says: "Concerning what is to happen, our people think that which they said Panaetius doubted, that in the end whole world will burn."[114] Seneca says: "with immense fires it will parch and burn mortals..... while all matter burns, all that is now in order will burn in a single fire."[115]

Now, scientifically the matter can be shown thus. Gradually, throughout the course of time, the motion of the stars can become slower, whereby finally the Moon will attract the Earth, and one star will collide with another; and thus the Earth, the Moon, the stars, and the Sun

---

[110] 2 Peter 3, 10: "Adveniet autem dies Domini ut fur, in quo coeli magno impetu transient, elementa vero calore solventur, terra autem et quae in ipsa sunt opera, exurentur."

[111] Hymn *Libera me*, from the Traditional Roman Ritual (Burial Service, Rite of Absolution): "Libera me, Domine, de morte aeterna, in die illa tremenda quando coeli et terra movendi sunt, dum veneris judicare saeculum per ignem." – ***The Translator.***

[112] OVID, *Metamorphoses* 1.254:

"Esse quoque in fatis reminiscitur adfore tempus,
Quo mare, quo tellus corruptaque regia coeli
Ardeat, et mundi moles operosa laboret."

[113] LUCAN, *Civil War*, Book 7:

"Hos, Caesar, populos si nunc non usseriit ignis
Uret cum terris, uret cum gurgite Ponti;
Communis mundo superest rogus."

[114] CICERO, *De natura deorum*, 2.46: "Eventurum nostri putant id de quo Panaetium addubitare dicebant, ut ad extremum omnis mundus ignesceret."

[115] SENECA, *Ad Marciam*: "Ignibus vastis torrebit, incendetque mortalia... omni fragante materia uno igne quidquid nunc est ex diposito ardebit."

will form one ardent mass. Otherwise, if natural causes are not sufficient, the divine power will intervene.

IV. – Third Conclusion: "The world will finally be renewed."

The Faith and the traditions of which we just spoke also bear witness to this, but there are arguments as well. It is clear that the world is not going to be annihilated. But because it is not going to be annihilated, it must have a congruent form and state. Therefore, the world, after it has lost its present state, will obtain a different congruent state. But a state congruent to the end of things cannot be chaos and horror; for God, who intends the consummation of all things, cannot allow them to end in disorder. Therefore, the world will not remain in a certain horror and disordering after the universal conflagration, but will be given both order and beauty.

Proof. The corporeal world is for the sake of man, that is, so that it serves him either for the sake of the sustenance of his corporeal life or for the sake of his advancement in divine knowledge. Therefore, it will be renovated so that it may serve man, if not for the sake of his corporeal life, at least for the sake of his advancement in divine knowledge, so that he may glimpse the divinity and the signs of the divine majesty in corporeal effects. Thus it is necessary that bodies receive an even greater influence from the divine goodness than now. Now, this influence consists in a certain splendor. For indeed, creatures lead to the knowledge of God by their beauty and comeliness. But the comeliness and beauty of bodies consists especially in light. Therefore, bodies will improve especially as far as their clarity. But their quantity and the mode of their improvement are known only to Him who will be the author of their improvement.[116]

It is also necessary that, in the last renewal, corruption and imperfect motion will cease to be. For since the renewal of the world will occur for the sake of man, it is necessary that it correspond to the renewal of man. But the renewed man will transition from a state of corruption to one of incorruption and perfect rest. Therefore, the world

---

[116] Cf. *Summa theologiae*, Supp., q. 91.

will be renewed in such a way that, all corruption having ceased, it will remain perpetually in a certain rest.[117]

What has been briefly explained is sufficient; for philosophy and science can only raise questions and offer uncertain arguments concerning the future state of the world.

[117] Cf. *Ibid.*

SECOND TREATISE

ON THE WORLD AS FAR AS ITS MATERIAL AND FORMAL CAUSES[1]

Knowing now the extrinsic principle *from which* (*unde*) the World comes to be, we must investigate the intrinsic principle *out of which* (*unde*) the World is. Now, the intrinsic causes out of which a thing is are called *constitutive* principles. Therefore, the whole question to be discussed now concerns the constitutive principles of bodies, both in general (*in communi*) and in particular (*in speciali*). Following this exposition we shall consider the properties of bodies.

FIRST QUESTION

[1] The following authors may be consulted: ARISTOTLE, *Physics*, *De generatione et corruptione*; PLATO, *Timaeus*; HENRI MARTIN, *Etudes sur le Timée*; LUCRETIUS, *De rerum natura*; PLUTARCH, *De placitis philos.*; J. SOURIS, *Théories naturalistes du monde et de la vie dans l'antiquité*; ST. AUGUSTINE, *Confessiones* 12; ST. THOMAS, *Commentaries on Aristotle*; JOHN OF ST. THOMAS, ALAMANNUS, COMPLUTENSES, CONIMBRICENSES, GUERINOIS, MAILHAT, GOUDIN, in their *Physics*; TOLEDO, *In Phys.*; SUÁREZ, *Disputationes Metaphysicae*; DESCARTES, *Les principes de la philosophie*; LEIBNITZ, *Système nouveau de la nature*; BOSCHOWICH, *Theoria Phil. natur.*; SECCHI, *L'unité des forces*; ZIGLIARA, *Summa Phil.* and *De mente conc. Vien.*; CARBONELLE, *Les confins de la science...*; LIBERATORE, *Philos. et De composito humano*; PESCH, *Phil. natu.*; DE SAN, *Cosmol.*; MIELLE, *De substantiae corporalis vi et ratione*; HENRI MARTIN, *Philos. spiritualiste de la nature*; FRÉDAULT, *Forme et Matière*; WURTZ, *La théorie atomique*; DE VORGES, *La constitution de l'être*; FARGES, *Matière et Forme*; FRANCK, *Dictionnaire Philos.*; BARTHÉLEMY SAINT-HILAIRE, *Préface de la traduction de la Physique d'Aristote*; CH. LEVÊQUE, *La Physique d'Aristote et la Science contemporaine*; NOURRISSON, *De l'Idée de matière*; DAURIAC, *Des notions de matière et de force dans les science de la nature*; DE MUNNYNCK, *Notes sur l'hylémorphisme*; NYS, *Le problème cosmologique, Cosmologie*, *Revue néo-scolastique*, 1904; P. DUHEM, *Le Mixte et la Combinaison chimique*; J. GREDT, *Elementa philosophiae*, t. 1, 3rd Ed.

On the constitutive principles of bodies in general

FIRST ARTICLE

The conditions of the first principles of bodies

I. – THE NOTION OF PRINCIPLE. *A principle* designates that from which something proceeds in any way; it signifies only an order of one thing to another, and it abstracts from the influx of one thing into the *esse* of another. Now, that which originates from a principle is called a *principiatum* [literally, a 'principled' thing]. But the notion of cause adds to this notion the note of a real influx of one thing into the *esse* of another, which is called the *effect* or the caused [thing]. And so, the name of cause implies a *dependence* of one thing upon another, but this is not the case with the principle: for this reason all causes are necessarily principles, but not every principle is a cause. In the present discussion, however, we take both terms, principle and cause, as meaning the same thing. We shall discuss their distinction in the treatise on Ontology.

II. – CONCLUSION: "The first principles of things are rightly defined by Aristotle: *Those things that are not from others, nor from each other, but all things are from them.*"[2]

For the absolutely first principles must be first in such a way that they are in no way *principiata*. Otherwise, if they were from others, they would require a principle; hence they would be *principiata*. Thus, if one would contribute to the constituting of the other, then neither of the two would be first, but each would be related to the one that constitutes it as a *principiatum* relates to its principle. Just as in division the members must not include each other, so in things the principles must not be from each other. Now, although causes, as causes, can be causes of each other, because they can have a mutual dependency whereby one flows into the *esse* of the other, first principles, nonetheless, can in no way be principles of each other. For if one thing is a principle, then the other is *principiatum* and already loses the *ratio*

[2] *Phys.* 1.6; ST. THOMAS, *In I Phys.*, lect. 10 (text 52).

of first principle. It is necessary, therefore, that the first principles be contraries like the members of a division.

Finally, it is required that all things be from them. If only some bodies were from these first principles, and not all things, then they are to be called principles *of some* bodies in particular, and not the principles of *bodies in general*.

III. – Corollaries. From this definition we can derive the following: (1) The first principles are not complete substances. For principles are those things from which a body coalesces as an essence that is *per se* one. But a complete substance is not *per se* one, but *per accidens* one; at present it is sufficient to remember this, for our discussion will return to this point many times. (2) The first principles are not *elements*; for 'element' designates a part that intrinsically composes that which is made out of others. Thus oxygen, hydrogen, nitrogen can be said to be the elements of some body, but not its principles, because it is further to be inquired by means of which things oxygen, hydrogen, and nitrogen come together.

IV. – Second Conclusion: "First principles must be contrary in some respect; indeed, they must be first contraries."

They are not contraries insofar as they *in fact are*, or insofar as they are in the composite, for it is thus that they are joined in one substance; but they are contraries *in becoming*. For it was said that they do not come to be from each other, but are rather opposed to each other as the members of a division. For coming to be is a kind of motion. But all motion implies a certain contrariety, namely, the coming to be of one and the ceasing to be of the other. Therefore, principles of becoming require a certain mutual contrariety.

Indeed, these principles must be the first contraries. First contrariety can be understood in three ways. First of all, we call first contraries the supreme differences that first divide a genus. Thus, Aristotle presents white and black, which first divide the genus of color, as first contraries; the other intermediary colors are second contraries. Secondly, we call first contrariety that which is found in the genus of substance, whatever it may be. Substance, of course, is the first genus among the predicaments. Therefore, the contrariety that affects substances can be said to be first contrariety.

From this one can argue thus: The constitutive principles of bodies must be substantial, because they compose the substance intrinsically. But substantial things are the first genus. Therefore, the principles of bodies are referred to the first genus. But the contrariety that is found in the first genus is said to be first contrariety. Therefore, the contrariety that is found among the principles of bodies is first contrariety.

Thirdly, and most especially, first contrariety exists between those things that are opposed to each other privatively. For first contrariety is wherever there is a principle of contrariety. But there is a principle of contrariety in privation. Therefore, those things that are privatively opposed to each other are first contraries.

That privation is a principle of contrariety is evident in itself, because all contrariety implies a privation, as vice includes the privation of virtue and hatred the privation of love. Now the principles of becoming in bodies are said first contraries primarily in this third way, namely, privatively; for becoming and motion imply the coming to be of one and the ceasing to be of another. The very definition of Aristotle implies that the principles of bodies, because they are contraries, are first contraries, because they do not come to be from other prior contraries, or from each other, but all things originate from them.

Those are the conditions assigned by the ancients, which for our times can be expressed in different words, thus:

V. – Third Conclusion: "The first principles of bodies must be such that by means of these principles the duality and the antinomies that science professes to find in bodies can be explained."

It is a well known fact of experience that some bodies are almost subject to contradictions: (1) They are passive and nonetheless produce activity and energy. (2) They enjoy unity and nonetheless are divisible and possess multiplicity. (3) In them there is something common and generic but also something proper and specific. (4) In them there is something that is always permanent but also something transient: when the tree becomes ash, the tree does not endure, but it is also not reduced to nothing.

The natural sciences have proven that matter and energy is always conserved indefinitely.[3] The law proposed by Lavoisier [the Law of the Conservation of Mass] is that the quantity of matter remains after any

---

[3] Cf. Treatise I, q. 1, a. 2, n. 11.

change or combination, which is clearly established from the sameness in the weight [before and after the change]. Hence, the dictum is very true: "Nothing is created; nothing disappears."

Thus [Newtonian] mechanics show that the sum of energy or power [in the universe] is always the same, such that, when the quantity of some motion seems to diminish, it is found again under the form of heat, and vice versa (even if some energy is lost): *the mechanical equivalent of heat.* There is, therefore, something permanent in bodies.

Also, chemistry teaches that properties vary, for they distinguish between mixtures and compounds. While in mixtures the same properties remain [as in the separated elements], in a compound a new composite emerges that possesses new properties. These specifically diverse properties are more clearly manifest in the generation of living things. There is, therefore, something in bodies that is transient and varies.

From these facts it follows that in bodies there is a certain duality; it cannot be, of course, that opposites and contradictories follow from the same thing. Therefore, in order to assign a congruent reason for these facts, we say that the first principles must be related in such a way that one is the root of passivity and inertia, of multiplicity and division, a common, generic, and permanent principle; and such that the other is the principle of activity and unity, a proper and specifying principle. – The passive and common principle, we shall call the 'potential', or 'material' principle; whereas we shall name the active and specifying principle the 'dynamic', or 'formal' principle.

VI. – A Problem to be Solved. There is no trouble in admitting that there are two principles in general, namely, one material and the other formal, abstracting from the nature of each; the difficulty is in determining their nature and their mutual relationships. Here is where philosophers disagree. Some deny the dynamic principle so that they may save the material principle; hence atomism. Others, in order to defend the dynamic principle, eliminate the material principle; hence dynamism. Finally, others defend the rightful place of both principles in such a way, however, that the material principle is related to the dynamic and formal principle as substantial potency to substantial act; hence hylemorphism, the scholastic system of prime matter and substantial form.

VII. – What These Three Systems Propose in Common. Atomism and dynamism agree on four things: (1) The constitutive principles of bodies are complete, elementary substances that are immutably permanent in their *esse*; for, whether they are atoms, or monads or powers, they are conceived as substances in act. (2) Consequently, bodies are merely aggregates of extended or simple substances. (3) In the world there is no generation or corruption, properly speaking, no substantial change, because constitutive elements have and retain immutable *esse*. (4) Bodies only differ accidentally among themselves, by reason of motion, place, disposition, order, powers.

Hylemorphism professes together with atomism that matter and extended substances exist, and with dynamism it defends the existence of an active and dynamic principle, but it differs from both of these theories on four points. It establishes that (1) the intrinsic principles of bodies are certainly substances, but they are incomplete; the two are related as potency and act; (2) bodies are not aggregates of many substances, but rather, something resulting from the union of their components that is *per se* one; (3) in the world there is substantial generation, and substances are truly generated and corrupted; (4) bodies differ from each other specifically and essentially.

We shall consider each system singly.

Atomism explained and refuted

I. – The Atomism of the Ancients. According to the atomists, bodies are aggregates of atoms caused by the power of motion. Atoms, in turn, are the smallest particles, which, although extended, nonetheless remain indivisible. The smallest part of a simple body is called an *atom*, whereas the smallest part of a composite body is called a *molecule*.

This doctrine of corpuscles takes various forms. Among the Ionians there was an opinion concerning the "one mobile."[4] Thales of Miletus, aware that living things arise from a seed that has a humid nature, and that they conserve life by means of food that also has a humid nature, claimed that the principle of all bodies is water. Anaximenes claimed the opposite; establishing that generation occurs from the change in the air and that life is conserved through the breathing of air, he held that air is the first principle. Heraclitus, however, having experienced the influence of fire in the generation of bodies and in the breathing of living things, established fire as the first principle. Empedocles, who is credited for the theory of the four elements (earth, water, fire, air), taught that the composition of bodies occurs through the mixture and separation of those elements. But he said that, for this to occur, certain empty intervals, or pores, are necessary, into which full particles, or atoms, can penetrate. To this doctrine, Anaxagoras added the theory of infinite number, or that of *homeomers*, that is, of atoms, which are infinitely small and infinitely numerous, homogeneous particles. These particles constitute all substances; hence in each of these atoms is found, as in seed form, something belonging to all substances; thus it is that "everything is the seed of everything," as they used to say.

After these men, Leucippus and Democritus refined the atomistic doctrine. The void and the *plenum* (the 'full') would be the principles of things, which the Eleatics denied. The *plenum* and the void are opposites; but, because the void lies between them, it is necessary that the *plenum* be discrete. It is on account of their physical indivisibility that atoms take their name. Bodies, therefore, are aggregates of atoms.

---

[4] Cf. St. Thomas, *In I Metaph.*, lect. 3.

Further, atoms are infinite in number, and they only differ among themselves in shape, place, or order, as A differs from N in shape, AN differs from NA in order, and N differs from Z in position. Epicurus made atomism famous, and Lucretius explained it and defended it in verse; the following two verses are well known: "All nature then, as it exists through itself, consists in two things: for there are bodies and the void."[5]

II. – THE ATOMISM OF THE MODERNS. The fate of atomism was that, after Lucretius, it would gradually vanish. But in the Sixteenth Century, the study of antiquity having been restored, the opinions of the atomists were revived. The first defender was Telesio (1508-1588); after him came Berigard (1578-1667). In the Seventeenth Century, Maignan, of the Order of the Minims, Sennert, and Magnen, in his work, *Democtritus Reviviscens,* defended atomism; but it is especially Gassendi and Descartes who made it famous. Gassendi restored the substance of the system of Democritus but rejected the idea that atoms are eternal and not produced. Descartes added many things to this opinion. He thinks that the essence of bodies consists in having three dimensions. But the mass of matter, which was created by God, had been separated into molecules, and three elements were produced out of the mutual collision of molecules. The first element is a very tenuous matter, like a most subtle dust; from this matter, the Sun and the fixed stars were formed. The second element has the form of globules or spheres that were produced by virtue of the rubbing off of angles; the heavens were thus established. The third element consists in triangular molecules that are like minute columns hollowed out spirally by three grooves in the manner of a snail shell: "they resemble small columns ribbed by three grooves or channels, and turned like a snail shell."[6]

Now, all things in bodies, even specific variety, are reduced to mechanical motion. Robert Desgabets, O.S.B., embraces and defends the opinion of Descartes.[7]

Atomism is also attributed to Newton, Muschenbrock, and Euler.

---

[5] *De rerum natura*, 1.419-420:
"Omnis, ut est igitur per se natura duabus
Consistit in rebus: nam corpora sunt et inane."

[6] *Principes de la Philosophie*, 3e partie, n. 90.

[7] Cf. P. LEMAIRE, *Dom Robert Desgabets, son système, son influence, son école*.

III. -- Atomism in Our Age. Dalton, an English chemist, applied the opinion of the atomists to the physical sciences, especially to explain chemical compounds, which occur according to determinate proportions and in which matter to a certain degree appears to be reducible to its smallest parts. Later scientists unanimously embraced this hypothesis. According to certain moderns, simple bodies, of course, consist of atoms of diverse species; but composite bodies are only additions and juxtapositions of simple elements that are immobile and permanent in their substantial *esse*. In this juxtaposition the molecules do not touch each other immediately, but a certain interval mediates between them, which is filled with a tenuous and weightless matter, commonly called *ether*. The ancients asserted that atoms moved like a whirlpool and that they congregated by means of motion and thus bodies were formed; according to the more recent thinkers, the atoms of the ether are in perpetual commotion, and through perpetual vibrations they generate all phenomena, namely, gravity, light, heat, adhesion, cohesion. Chemical *adhesion* is the power that joins heterogeneous atoms; but *cohesion* is the power that connects homogeneous molecules.

The aforesaid authors, therefore, hold that there is no true change in the chemical compounds by means of which bodies are produced, but only diverse vibrations that produce diverse impressions in the senses.

Among the modern atomists, some defend a merely *mechanical* atomism, that is, they explain all things through matter and merely mechanical motion; others defend a *chemical* atomism, which consists in the principles of chemistry, according to what they say about simple and compound bodies. And these disagree with the mechanists if they think that the powers that act upon bodies cannot be recalled to locomotion, but that they are distinct from motion and that they emanate from the essence of atoms. If they reject this, then chemical atomism would not really differ from mechanical atomism. For this reason, it is not necessary to distinguish these two forms so sedulously.

Mechanical atomism was especially defended by P. Secchi, in his work, *L'unité des forces physiques*; and chemical atomism by Tongiorgi, Panciani, Bottalla and many others. It must be noted, however, that these authors do not sufficiently distinguish scientific atomism from philosophical atomism.

IV. First Conclusion: "Even if it is admitted that the matter that is in mixtures and compounds is to a certain degree reducible to the smallest particles, it cannot be conceded that atoms are separated all around by intervals; nor does this prove that atoms are the quidditative principles of bodies."

Proof of the First Part. The facts and laws of chemistry prove that matter is reducible to a certain degree:

First, the Law of Definite Proportions (Proust's Law). For two bodies to coalesce into one chemical compound, not just any weight or mass is sufficient, but there is always a definite and invariable proportion of weight and mass according to which bodies are associated. And this shows that bodies are reducible to the smallest things, not indeterminately, but to a certain degree.

Second, the Law of Multiple Proportions (Dalton's Law) proves this. When two diverse bodies become a compound, it is always required that their number be a multiple of the prior proportions: for example, in a first compound there are fourteen parts of nitrogen and eight of oxygen; in a second compound there will be sixteen parts of oxygen; in a third, twenty four; in a fourth, thirty two. Therefore, the reducing of matter occurs to a determinate degree.

Third, the Law of Equivalent Weights (discovered by Wenzel and Richter). When one element replaces another, the substitution does not occur in just any proportion, but in that proportion under which those elements can combine chemically. We admit, therefore, that matter can be dissolved to a certain degree, and the ancient Scholastics did not deny this, as their axiom indicates: "Bodies only act in a dissolved state" (*Corpora non agunt nisi soluta*); they, moreover, called the smallest parts to which matter is reducible, "the smallest elementary things" (*minima elementaria*).

Proof of the Second Part. Chemical facts show one thing: atoms are divided at the instant in which they are combined or dissociated and, hence, one cannot conclude that atoms remain joined in the compound state. We do not deny that there is some intermediary interruption between them; but it is altogether necessary that at least in some way they unite so that they form one body, or one continuum that is actually undivided, even if it is divisible in potency. For if atoms are in no way united, but are separated by empty intervals, then we would have to admit action at a distance. – *The major premise is clear.* For those things that are separated by empty space do not touch each other.

Therefore, there is a distance between them. Therefore, action through empty space is action at a distance. *The minor premise* shall be demonstrated in Metaphysics but is briefly proven here. Since action follows being, the agent is in no way able to act where it is not. But the agent is not in something distant, as is evident. Therefore, it cannot act at a distance; therefore, action at a distance is contradictory.

It is again proven by refuting the arguments of adversaries. They say that there are empty intervals in order to save the law of magnitude proposed by Gay-Lussac so that vibratory motion may be admitted in atoms and molecules. We respond: All these things can be sufficiently explained if some property be attributed to bodies by reason of which a large quantity of matter can be present in small dimensions, and the same atoms can occupy smaller or greater spaces, and thus be expanded and compressed according to the density or rarity, as the Scholastics supposed, or according to elasticity and compressibility, as others say. Having admitted porosity, there is nothing that evinces that atoms fly about through empty space.

Now we refute others who contend that atoms move in ether. For ether, no matter how subtle it is supposed to be, is nonetheless a corporeal *ens*. Therefore, ether either has continuous extension or exists as separate atoms. If it has continuous extension, then we must admit the existence of the continuum, as the Scholastics do, and this is what we intend. But if it exists as separate atoms, then we argue thus: These atoms either move in an empty space, and so incur all the difficulties that we have hitherto exposed; or they move in some other, subtler ether, and this ether will move in another, yet subtler ether, and thus we have an infinite regression. Therefore, even though we may admit the existence of atoms, nevertheless it should not be admitted that they are separate in a empty space or flying about in ether.

Further, many scientists attack the reality of atoms. Huxley says: "I shall believe in the existence of atoms under the condition that they first prove the existence of those atoms."[8] And Frédault says: "There is no learned man today who does not consider atoms as an idea of the mind that one uses to express the conditional relationships of quantity and equivalence in chemistry.... That is a word that has its utility to

[8] *Scienza italiana*, v. 2, p. 117: "Je croirai à l'existence des atomes à la condition que l'on me prouve d'abord l'existence de ces atomes."

unify the equations of movement; and, if it has its dangers, it is necessary to see that it is nonetheless only a word."[9]

There is another word that is now celebrated among scientists: *ions*. Perhaps the question of the constitution of bodies will be illustrated by means of these particles that have been discovered so recently. For this reason we applaud those learned men who work in the study of ions.

Proof of the Third Part. It is absurd to posit as one of the constitutive parts of a body the constituted body as a whole. But atoms are already constituted bodies. Therefore, it is absurd to posit atoms as quidditative or constitutive parts of bodies. *The minor premise is per se known.* Atoms are, of course, of very tiny and minute quantity; they are corpuscles, but they possess the whole quiddity of a body. They can, therefore, at most said to be the elements of bodies, but in no way the first principles of bodies.

V. – SECOND CONCLUSION: "Atomism, taken philosophically, explains the *ratio* of neither composite bodies nor simple bodies."[10] This is the opinion of the more recent Scholastics.

First of all, we suppose the existence of simple and composite bodies as an unshakable fact. Simple bodies can be defined thus: Those things that are reducible to others, which themselves are not reducible to anything else; these are about seventy in number. Composite bodies result out of the combination of simple bodies.

---

[9] *Forme et matière*, p. 100: "Il n'est pas un savant aujourd'hui qui ne considère l'atome comme une vue de l'esprit dont on se sert pour exprimer des rapports conditionnels de quantité et d'équivalence en chimie.... C'est un mot qui a son utilité pour réduire à l'unité des équations de mouvement; et, s'il a ses dangers, il faut bien voir qu'il n'est cependant qu'un mot."

[10] Our conclusions and arguments refute only *philosophical* atomism insofar as it asserts that atoms are in a metaphysical sense the essential principles of bodies. But they concede to physicists and chemists that atoms can be called the principles of bodies in a scientific sense, that is, proximate principles, or elements, of which science only treats.

Hence the opposition between scientists and Scholastics must vanish. For scientists consider the matter physically; hence they may call atoms physical principles if they wish, and they may enjoy complete liberty in the ambit of the proper object of their study. But we speak directly in another order; for we speculate about this matter philosophically and metaphysically.

Therefore, we argue thus: It is certain that there are specifically-diverse simple bodies. But the opinion of the atomists cannot provide any argument for this specific diversity. For some contend that specific diversity comes from the diversity of shapes. But shape is an accident; it presupposes, therefore, an already constituted body. Further, it would have to be asked where the diversity of shape comes from. Others reply that the cause of specific diversity is the divine will. This is no response, for beyond the divine will, which is the first cause, we must also assign a proximate cause: for indeed, in every natural thing is infused some intrinsic principle. Certain thinkers think that it is produced from the vibration of ether, but this is not the case. For the species of things are invariable and immutable. But the vibrations of ether, even in the same place, change and vary perpetually. Therefore, such vibrations cannot produce diverse species. Moreover, atoms are either of diverse natures or of the same nature. If of the same nature, there is no reason why the vibrations of ether act upon one atom in one way and upon the other in another way, or why it forms these atoms into a compound and not some other atoms. But if atoms are of diverse species, we are now begging the question (*petitio principii*), and again the atomists would have to ask whence the specific diversity of atoms originates. One can always object against the atomists what we already pointed out: Both atoms and ether are constituted bodies. Therefore, it must be asked whence proceeds the intrinsic constitution of these bodies. Atomism, therefore, does not give an account of simple bodies.

Nor does it give an account of composite bodies. Chemical compounds and simple mixtures differ in two respects: for, first of all, a body that arises out of a chemical compound is altogether homogeneous, as all the parts of the compound between oxygen and hydrogen are water. Second, a body that is a chemical compound exhibits properties that are specifically diverse from those of the elements of which it is composed. And each of these, if true, contradicts atomism. *Therefore.* The first is true, for according to the atomists, elements are only juxtaposed because atoms remain the same as before. But, if the atoms of oxygen and hydrogen are juxtaposed, then water is not a homogeneous body, and not all the parts of water will be water; but one part will be oxygen and the other hydrogen. The second is also true. For if the merely juxtaposed elements remained, there could be a union, an adding of powers and properties, but this addition does not constitute specifically diverse properties, as when barley and corn are mixed, they make a greater sum, but not a new species.[11]

VI. – Third Conclusion: "Philosophical atomism results in many difficulties." Atomism, taken philosophically, is incompatible with the notion of beauty in the universe, with the notion of substance, and with the notion of life; it robs art and modern technology from their object and reality, and finally leads to skepticism. We prove this in parts:

(1) The beauty of the Universe consists in the specific variety of things and in the essentially diverse degrees in which beings participate in the divine goodness. But matter and motion, since they are common to all, cannot establish specific variety and essential diversity. Therefore, if beyond matter and motion no other formal and distinctive principle is admitted, then the beauty of the Universe is eliminated. But atomism is incompatible with this formal principle. Therefore, it is incompatible with the notion of beauty. Further, in the preceding number we showed why atomism cannot assign a cause of the specific diversity of bodies.

(2) Substance consists in the notion of *ens per se*. But in the opinion of these adversaries, corporeal substances would consist in the notion of *ens per accidens*. Therefore, atomism rejects the notion of substance. *Proof of the Minor.* In the foregoing hypothesis, the constitutive principles of a substance would be, on the one hand, atomistic matter, and on the other hand, motion, number, or the disposition of atoms. But, since motion, disposition, and number are accidents, a composite that is made out of matter, motion, and disposition will be an accidental thing, or an *ens per accidens*. Therefore, the notion of corporeal substance would have to be replaced by the notion of an *ens per accidens*. *Proof.* A thing that is *per se* one does not result out of many beings in act. But each atom is an *ens* in act. Therefore, if a body is put together out of atoms as its first principles, then it will not be something that is *per se* one, and so it will not be a substance. The major premise is an axiom. For every *ens* in act has *esse*, *simpliciter*, or is essentially perfect. But out of two perfect things, for example, out of two joined stones, does not arise something that is *per se* one. Therefore, out of two beings in act does not come to be

[11] See below, q. 2, a. 4, on the permanence of elements in a mixture.

something that is *per se* one. The minor premise is admitted by the adversaries. For atoms are conceived as certain indivisible corpuscles that exist in act. – And so does atomism destroy the notion of substance.

(3) It belongs to the notion of a living thing that it can move from within. But, if we admit atomism, there is no motion from within. Therefore, it rejects the notion of a living thing. *Proof of the Minor.* Motion from within requires that *the same* substance move itself according to diverse parts. But in atomism it cannot be the case that one and the same substance move itself according to diverse parts, but rather, there is an *aggregate* of many substances of which one is moved by the other. *Therefore.* For in the opinion that we are refuting, the living thing is an aggregate of atoms of which some move others. But every atom is a certain substance discrete and distinct from other atoms. Therefore, there cannot be one substance that moves itself, but many substances of which some are moved by others.

(4) It robs art and modern technology of its object and reality. ***Proof.*** Art and modern technology work wonders, especially things concerning physics. Now, atomism robs physics of its object and reality. Therefore, it robs art and industry of its object and reality. *Proof of the Minor.* The object and reality of physics presupposes locomotion, and the real transition from one space to another. But if we admit atomism, locomotion is not possible, nor is the real transition from one space to another. *Therefore.* For locomotion requires something mobile *per se* that has real parts by virtue of which it can be partially at the terminus from which (*terminus a quo*) the motion begins and partially at the terminus to which (*terminus ad quem*) it tends. Further, atoms, insofar as they are indivisible, do not have parts by virtue of which the mobile thing can be partly at the *terminus a quo* and partly at the *terminus ad quem*. Therefore, motion is impossible.

Further, the real transition from one space to another requires a continuum; for if all things are reduced to one point, then the mobile thing always remains in the same point. But atomism destroys the continuum and reduces all things to a single point. *Therefore. Proof of the Minor.* We either suppose that atoms have distance between them or that they touch each other. If there is distance between them, then there is no continuum, as is clear. But if they touch each other, since they are indivisible, they touch each other entirely. But those things that touch each other entirely do not form a continuum, because they do not have extremes that are distinct from their center. Therefore, atoms cannot form a continuum; hence all things are reduced to a point.

Atomists, therefore, who boast of art and modern technology, are found to oppose art and industry by their very system.

(5) If all things are explained by vibrations and motion, then the senses err when they report the intrinsic qualities of bodies, their internal activities, and their specific diversity. It also follows that all sensible cognitions are nothing other than certain reactions of human sensibility; and hence the subject only knows his impressions, or his subjective phenomena, and does not know with certainty whether there is something objective really corresponds to these subjective impressions. But this is phenomenalism, which prepares the way for skepticism. This is why many atomists, both among the ancients—like Protagoras, Pyrrhus, and Epicurus—and among the moderns, have fallen into skepticism.[12]

VII. – Difficulties Resolved.

1st Objection. Science proves that there is porosity in bodies. But porosity can only be explained by discontinuous atoms. Therefore, there are discontinuous atoms. ***Reply***. *I distinguish the minor.* That porosity can only be explained by atoms that are discontinuous *in some respect*, I concede; but that it can only be explained by atoms that are discontinuous *in all respects*, I deny. *I distinguish the conclusion.* That there are atoms that are discontinuous in all respects, I deny; but that there are atoms that are discontinuous in some respect, and that are not the constitutive principles of bodies, I concede.

It has been shown in the first conclusion that the absolute discontinuity of atoms is contradictory; elsewhere we admitted a certain interruption or discontinuity among atoms in some respect. But this is quite sufficient for explaining porosity. Now, even if we admit the existence of atoms, it does not follow that they are the first constitutive principles of bodies, but only the smallest elementary particles, as we suggested earlier.

2nd Objection. Expansion and compression require a distance between atoms. Therefore, there is a distance between atoms. ***Reply***.

[12] Again we remind the reader that our arguments are addressed only against philosophical atomism. For these difficulties do not arise if atoms are taken in the scientific sense, not as essential principles, but as principles that are the proper object of the physical sciences.

Rarity and density require at the most a certain interruption, or discontinuity, in some respect, as was said; but they do not at all require an absolute distance or empty space; for these facts are excellently explained in the scholastic doctrine on potency and act. Of course, bodies, which have certain dimensions in act, are in potency to having different dimensions; and thus they are said to expand (or contract) when they transition from having greater (or lesser) dimensions in potency to having them in act.

3rd Objection. The principles of extended bodies must be extended. But the extended principles are atoms. Therefore, atoms are the principles of bodies. ***Reply***. *I distinguish the major.* That the principles of extended bodies must be extended, that is, that they are the root whence extension is obtained, I concede; but that they can be extended formally and in act, I deny. *I contradistinguish the minor.* [That atoms must be extended, that is, that they are the root whence extension is obtained, I concede; but that they can be extended formally and in act, I deny.] *I deny the objection's conclusion.* – The principles of bodies must not be extended formally and in act: otherwise they would already be constituted bodies, for actual extension is an accident that presupposes a constituted substance. Hence, atoms, because they are extended, although indivisible, already fall under the *ratio* of bodies and cannot have the properties of first principles. We shall weigh the other difficulties in our defense of hylemorphism.[13]

----

[13] Those who seek an examination of scientific atomism may consult D. NYS, *Cosmologie*, L. I.

Dynamism explained and refuted

I. – Ancient Dynamism. Dynamism, in order to save the specific and active principle in a corporeal substance, drives away the material principle, or the principle of extension, and asserts that simple substances, or simple powers, are the intrinsic constitutive principles of bodies. Pythagoras is referred to by some as the first father of dynamism due to his theory of number. This was Pythagoras' thinking process. We observe in things, whether physical or voluntary, a certain equality; thus health and justice exhibit a certain property of equality. But equality pertains primarily to number. Therefore, numbers are the principles of things. Again, we see that musical harmonies and proportions are produced by the nature of numbers. Hence, it is permitted to conclude that numbers are the principles of things, and the very heavens are a certain nature and harmony of numbers.[14]

Pythagoras posited two essential principles: the *monad* and the *dyad*. Now, it is not clear whether monads are in themselves unextended principles or not. Aristotle, however, reports that the Pythagoreans acknowledged in monads a certain magnitude or extension. But if monads are extended, then they are not really distinct from atoms, from which Ravaisson infers that the teaching of Pythagoras can be called a mathematical version of atomism.

Plato, who taught that natures of sensible and particular things are constituted through the participation of the ideas, or of separate forms that subsist *per se*, is considered to be the precursor of the dynamists.

II. – Modern Dynamism. Whatever may be the case concerning the opinions of Pythagoras and Plato, it is certain that dynamism became famous only since the time of Leibnitz. Leibnitz playfully imagined that bodies consist of simple elements, which he named *monads*. In order to refute Descartes, who claimed that the essence of bodies consists in extension and passivity, he established that substance necessarily involves simplicity and activity; indeed, he defined substance as an *ens* endowed with the power to act. Monads, therefore, must be principles

[14] Cf. St. Thomas, *In I Metaph.*, lect. 7.

that are both simple and active, and, although they do not in themselves have extension, extension nonetheless results from their position, because extension is nothing other than the simultaneous and continuous repetition of position. Monads are also infinite in number, and dissimilar among themselves; they do not occupy any space, nor does is there empty space between them; they enjoy a certain perception and a certain appetite by which they unite among themselves.

Leibnitz's system had many followers, who nonetheless changed many things in their master's doctrine. Wolff accepted almost the whole system, except that he replaced the perception and appetite of monads with a motive power. Fr. Boschovich, S.J., a professor of the Roman College, disagreed with Leibnitz on many points. Mondads, according to him, are not endowed with perception and appetite, but with attractive and repulsive powers; hence the name dynamism (*dynamis* = power). – Through the attractive powers the body becomes one, but through the repulsive powers bodies prevent mutual compenetration and being reduced to a point. Further, monads are finite in number, homogeneous, and between them there is an interval of empty space, which can be indefinitely increased or decreased.

Beyond the mechanic power of attraction and repulsion, the idealists, following Kant, admitted certain subsistent intrinsic powers which they posit as the first elements of bodies. These powers, which are called "plastic," are directed by their nature to a certain and determinate *end*. For this reason, this theory is called *teleological dynamism* (*telos* = end). Kant, however, did not intend to define what the constitution of bodies is *in itself*; but only how it is *conceived* by us. Now, body is conceived as an aggregate of simple powers: hence, it is called *idealistic* dynamism.

III. – Dynamism in Our Age. Among those who favor dynamism are Maine de Biran, Dugald-Stewart, De Lammennais, Moigno, Ubaghs, and many others, both among scientists and among Catholics who think that this way the union between body and soul and the mystery of the most holy Eucharist can be explained more easily. Many adherents of the system of Fr. Boschovich claim that points are inflated, insofar as these simple beings, although formally unextended, are nonetheless virtually extended, and thus they can touch each other without penetrating each other.

IV. – CONCLUSION: "The dynamist system can in nowise be proven."

1st Argument. Common sense and experience attests to the fact that bodies are extended. But dynamism is incompatible with objective extension. *Therefore. Proof of the Minor.* The monads that Leibnitz posits, and the simple powers that the other dynamists posit, are simple and unextended beings. But, while many unextended beings can certainly produce a number, they can no wise produce magnitude or extension; for, of course, the unextended added onto the unextended cannot produce something extended.

One cannot respond that monads are virtually extended, in the way that the soul, although simple, is commensurate to the divisible parts of space. For (1) the parts of real space presuppose an extended body. Therefore, since points are commensurate to the parts of space and are virtually extended, they require an already-constituted body; they do not constitute it. And (2) many souls joined to each other would never make something formally extended. Therefore, neither can these points joined together bring about formal extension.

2nd Argument. Monads and those simple beings either touch each other or have distance among them; there is no middle ground. They do not touch each other, because contact requires an extended surface, but the aforesaid beings, insofar as they are simple, neither receive something extended nor can constitute formal extension, as we have already shown. Otherwise, if they touched each other, they would have to touch each other entirely, because they do not admit of parts. But those things that touch each other entirely are in the same place in space. Therefore, all points would be in the same part of space. But if monads have distance among them, then we must admit that there is empty space among them, in which case they will not be able to act upon each other, if it is indeed established that there is not action at a distance or action through empty space.

3rd Argument. Dynamism, in whatever form it is considered, is incompatible with the unity of body. All dynamists agree on this, that all bodies are made out of many beings that are complete in themselves. But something that is *per se* one does not come about from many complete beings. Therefore, in the dynamist system a body is not *per se* one. *Explanation of the Minor.* Unity of aggregation or of order can always come about from many beings; but, in order that a substantial unity come about, it is necessary that the components change, or that they relate mutually as potency to act. Further, the simple beings that the dynamists posit neither change nor are related to each other as potency to act. *Therefore.*

4th Argument (against those who posit simple powers).

Those powers are either something subsistent or operative potencies that exist in, and emanate from, some subject. But both hypotheses run into difficulties. *Therefore. Proof of the Minor.* A *per se* subsistent power is an immediately operative substance. But it is contradictory for a created substance to be immediately operative. Therefore, a subsistent power is contradictory. The minor proof is proven in the treatise on causes. This proof is briefly formulated thus: Since potency and act are in the same genus, an operative potency must be in the same genus as operation. But created operation is in the genus of accident. Therefore, all operative potencies are accidents. Therefore, there is no created substance that is immediately operative.[15]

But if attractive, repulsive, and plastic powers are potencies that emanate from some subject, then they already are predicamental properties and accidents; consequently, they presuppose the constituted essence of a body. Thus, we must ask what this subject from which they emanate is. Is it not simple and subsistent? In this case, it must be called an intellectual substance. But if it is a composite being, then it is a contradiction to say that it is a simple being. Therefore, the other hypothesis is impossible.

5th Argument **(against those who posit simple points)**. Those points are either in the genus of quantity or not. But in either hypothesis it is impossible that they be the constitutive principles of bodies. *Therefore. Proof of the Minor.* If they are in the genus of quantity, then they are the termini of a body of continuous quantity. But the terminus of a body of continuous quantity clearly requires a body that is already constituted. Therefore, they cannot be constitutive principles.

But if they are outside of the genus of quantity, then they certainly are immaterial acts. But an immaterial act, although it can act and be in the body, it nonetheless cannot constitute an extended body; thus, all angels joined together can never form a corporeal substance. *Therefore.*

V. – DYNAMIC ATOMISM. It is established from the foregoing that the principles of bodies are not matter alone or powers alone, but

---

[15] Cf. *Metaphysics: Ontology*. Treatise IV, q. 2, a. 3.

two elements are required, namely, a material element and a dynamic element. Pure atomism and pure dynamism, therefore, are systems that ought to be rejected; but perhaps the truth is found in a certain combination of these systems, where atomism provides the material element and dynamism the dynamic element. The matter can be explained thus: Atoms consist of two principles: some extended reality, namely matter, and some intrinsic power that is especially manifested through resistance; in a word, the constitutive principles are *matter* and *power*. This is *dynamist atomism*, which is attributed to Newton and which is especially defended by Henry Martin, in his work: *Philosophie spiritualiste de la nature*.[16]

VI. – Explanation of the Preceding System. That intrinsic power is either a property that is accidental to the essence of the emanating atom, or is a principle that is substantial to the essence of the persisting atom. If the former, then we infer that the constitutive whole of the atom is matter or motion, and hence we fall into pure atomism: for atomism does not deny that there are powers in atoms, but denies that powers are the constitutive principles of atoms. If the former, then we fall back into the scholastic system, at least the substance of it. For then two principles are admitted: the one is a substantial principle out of which we get extension, and the other is a substantial principle out of which activity flows. But the substantial principle out of which extension flows is prime matter, and the substantial principle from which activity emanates is substantial form, as will be clear from what we shall say below. Therefore, this opinion posits prime matter and substantial form.

We said that the aforementioned system coincides with at least *the substance of* the scholastic system; for there are two points on which there could be disagreement. First, dynamist atomism seems to admit as many substantial forms as there are atoms, but we defend the idea that there is a single substantial form. Second, the substantial power is either a principle that is immediately operative or it only designates a radical principle from which operation emanates. If the latter is defended, then there is thorough agreement with the Thomists, who hold that form is not immediately operative, but a radical principle of operation. But if it is established that that substantial principle is immediately operative,

[16] Tom. I, 2e Part., ch. 8.

then we run into a great difficulty, because no created substance is immediately operative. It must be confessed, however, that Scholastics do not agree among themselves regarding these two points. Hence, it would not be an absolute obstacle for dynamist atomism to be reducible in substance to the scholastic system.[17]

VII. – Difficulties Resolved.

1st ***Objection***. First principles must be simple, because they come neither from other principles nor from each other. But monads, or the powers of the dynamists, are simple beings. Therefore, they are the first principles of bodies. ***Reply***. *I distinguish the major.* That they must be simple, i.e., they must be substantial parts that are not divisible into other elements, I concede; but that they must be simple substances or simple accidents, I deny. *I contradistinguish the minor.* That monads, or powers, are simple beings, i.e., substances or simple accidents, I concede; but that they are substantial parts in the sense explained, I deny. *And I deny the conclusion.*

First principles must not be divisible into others, otherwise they would be *principiata*; on the other hand, they must not be complete substances, otherwise they would not constitute something *per se* one. Nonetheless, since they are constitutive of a substance, they must be substantial parts. Now, monads are not substantial parts, but complete simple substances; and simple powers are not substantial parts, but accidents. Therefore, both monads and simple powers contradict the essential properties of first principles.

2nd ***Objection***. A body is a substance endowed with activity. But the constitutive principles of a substance that is endowed with activity must be active principles, or active powers. Therefore, powers are the constitutive principles of bodies. ***Reply***. *I distinguish the major.* That a body is a substance endowed with activity in the sense that activity and the operative power are properties that are inseparable from the substance, I concede; but that this is so in the sense that activity is the formal *ratio* of a substance, I deny. *I contradistinguish the minor.* That the constitutive principles of a substance whose formal *ratio* is activity must be themselves active, I concede; and also that the constitutive principles of a substance whose activity is only a property must be

[17] For a thorough treatment of dynamist atomism, see Nys, *Cosmol.*, L. IV.

themselves active, i.e., the first root whence activity and operation can be found, I concede; but that they must be immediately operative, I deny. *And I deny the conclusion.*

The dynamists rely upon a false foundation in replacing the constitutive principles of substance with activity, or with the power to act. The truth of the matter is that activity is a property that requires an already constituted substance, for acting follows being. Therefore, the principles of bodies must not be immediately operative powers; but it is sufficient that there be found in them something whence activity can be found. Further, in the scholastic system there is a substantial form which, although not immediately operative, is nonetheless a first principle of activity.

3rd Argument. Although the single monads do not constitute extension, nonetheless from their collection a continuum can be put together. Therefore, the arguments against dynamism that were offered above falter. ***Reply.*** *I deny the antecedent.* For the collection of simple beings certainly forms a number, but in no way does it form a continuum, just as, if all human souls are joined together, no extension would result there from. If no monad has order to extension, nor will a collection of monads be able to produce extension.

4th Argument. At in least dynamism one does not run into the difficulty of action at a distance. "We have abundantly anticipated—replies Boschovich—that which pertains to action at a distance, since it could come about that any point acts upon itself and in the direction of action and energy by another point; or that God, according to the free law established by Himself in making nature, produces motion in either point." ***Reply.*** In no way is it understood how one point can act upon another; for they either are distant from each other, and then we already have action at a distance, or they touch each other, and, since they are simple, they touch each other entirely and now all points are reduced to the same place in space, which is absurd. And if God Himself, as Boschovich would have it, produces motion in each point, then we would fall into the theory of *pre-established harmony*, or mere *occasionalism*, and then all true activity would be removed from bodies.

FOURTH ARTICLE

The existence of matter and form is proven

I. – The Scholastic System. This system is reduced to three headings: (1) There is in bodies a material substantial principle and a formal substantial principle; (2) each principle is an incomplete substance; (3) the material principle is related to the formal as potency to act. We shall show these three points in the present article; but the remaining points that regard the intimate nature of matter and form will be considered in the following question.

II. – First Conclusion: "In all natural bodies there is some material principle."

Proof. The powers that we experience in natural bodies presuppose some substantial principle that is the root of extension. But the principle that is the root of extension is material. Therefore, in all natural bodies there is some material substantial principle. *Proof of the Major.* The powers that act in bodies are accidents. But all accidents require a substantial substrate. Therefore, in all bodies it is necessary that there be found a substantial substrate. – Further, that principle is the root of extension. For the activity of bodies occurs in space and is diffused through space, and bodies act upon each other through contact. But space requires extension, and thus contact requires an extended surface. Therefore, the activity and powers of bodies presuppose a substantial principle that is the root of extension. *The minor is established.* 'Principle of extension' and 'material principle' are inseparable principles; by 'matter', of course, we understand that which is the root of quantity and extension. We ascertain, therefore, that in bodies there is some material substantial principle. Further, the conclusion is sufficiently established from the refutation of dynamism.

The *Law of Equivalent Weights* also proves this. After any change, the same weight always remains. But weight arises out of matter, for only matter is properly weighable; power and form are not. Therefore, there is matter that remains after a given change, and which therefore is said to be unproduced and incorruptible.

III. – Second Conclusion: "In every organic body there is found, beyond its matter, a formal, or dynamic, principle."

1st Argument (specifically for animals). There is in animals some principle that reduces to one and expresses in one representation those things that are dispersed and diffused in things: thus when the sheep sees the wolf, it gathers into one the many things that are dispersed and diffused in its exterior. But matter alone cannot represent all these things. Therefore, beyond matter there is in animals some principle that gathers diffused and dispersed things into one; and we call this the formal principle. *Proof of the Minor.* Matter is the principle of multiplicity and diffusion, whose parts have among themselves only a unity of continuity. But an *ens* that is the root and cause of continuity and whose parts retain only a unity of continuity cannot be the principle of such marvelous and vital unity such as we find in animals. *Therefore.*

2nd Argument* (generally for all organisms).** In every organic body we find some interior, immanent power that subordinates matter to itself and directs it, and that inclines the parts and the actions of the parts to the utility of the whole, as is evident in the plant, in which each part is ordered to the good of the whole. But the power or principle that directs matter cannot be matter, nor even the powers of matter. For indeed, the powers of matter have a transient action; even the activity of a material molecule is transient and is consumed when it is brought forth. They, therefore, cannot explain this internal and immanent tendency. Therefore, beyond matter, there is in organic bodies some formal principle that is superior to matter and that is really distinct from matter. ***Proof. A living thing always remains formally the same. But, even though the generic matter is the same, nonetheless the matter of individuals, as science attests, is in perpetual flux; such that after some years all the matter of a living thing is replaced. Therefore, the matter that flows, and the form by means of which the living thing remains always the same, are really distinct. This thesis will reappear in our philosophical treatise on Biology, where we shall show that the principle of life cannot be derived from matter or from the powers of matter.

IV. – Third Conclusion: "In all bodies, even inorganic bodies, there is a formal principle that is really distinct from matter."

1st Argument. In bodies there must be some principle by means of which the body becomes one *ens*, by means of which it is placed in some firm and permanent species, and consequently in a firm state of *esse* and acting; otherwise, all things would be subject to flux. But matter, since it is a principle of diffusion, cannot establish this perfect unity; since it is in perpetual flux, it cannot give a firm and permanent state of *esse*. Therefore, there must be in all bodies another principle distinct from matter.

In other words: physical laws can be stable only if one concedes that there are in things certain principles of acting consistently and uniformly that are proper to each thing. And one can only save the certitude of the physical sciences by acknowledging that there is in each body a certain internal power that determines the body to certain consistent effects that can be predicted with certainty.

But the internal power of acting consistently is the dynamic and formal principle.

Hence one can make an effective argument: We find in the world a certain perennial order and there are fixed and permanent laws by which inorganic nature is ruled, laws proper to each kind of thing. We already showed this in Treatise I, where we spoke of the twofold order, namely, the teleological order and the dynamic order, and we shall prove it again in Treatise III, where we will speak of the laws of nature. But order and invariable laws argue for an interior tendency and finality in things: for unless one attributes to individual things an interior finality, the course and order of nature cannot be established invariably, and the physical sciences will not be infallible. Therefore, there is in the inorganic world an internal tendency and finality proper to each body or to each nature. But an interior finality requires an interior and specifying principle from which the finality originates, and which we have named the formal and dynamic principle. Therefore, there is in each inorganic body a specifying and formal principle.[18]

2nd Argument. In natural bodies we find many antinomies and many opposite elements, for one and the same body is both permanent and transient, both diffused and one, both passive and active, etc., as we saw above, art. I, no. V. But those things that are opposed to each other, even though they may be derived from one *ens* according to diverse principles, cannot, however, be produced by one *ens* according to one and the same principle. Therefore, in the natural body we must admit two principles that are really distinct: one material, the other formal. – This argument can be proposed in a different way:

---

[18] This argument is implicitly given by NYS, *Cosmologie*, nn. 299ff.

Irreducible properties require irreducible substantial principles. But the principal properties of bodies, namely, quantity and quality, extension and power, are irreducible. Therefore, there are irreducible principles in bodies, of which one is the root of quantity, and this is matter and potency; and the other is the root of quality, and this is form. These are substantial principles, since they are the first root of accidental properties.

Proof of the Minor. The concepts of extension and quantity denote composition and divisibility; but the concept of power denotes something simple; for it is not subject to shape, dimensions and, although it may be in an extended body, in itself does not require extension, but act. Now, act, according to its own *ratio*, is simple. But the simple and the composite, the divisible and the indivisible, are irreducible. Therefore, extension and power, quantity and quality, are irreducible properties.

This argument shows that there is in bodies a substantial potency and a substantial act. Now, this brings us to our purpose: for else what do we understand by the term 'prime matter', if not a substantial potency, and what do we understand by the term 'form', if not a substantial act? The duality involved in bodies, therefore, shows that there are in them two substantial principles that are related in the manner of potency and act.

Of course, irreducible properties also require an internal finality: for unless there were a fixed law and an immanent finality from which it springs, the antinomy would not be always and uniformly found. Therefore the present argument is again confirmed by the preceding. The proof that uses irreducibility as a middle term, however, is not to be abandoned: it clearly shows of course that there is in bodies a substantial potency and a substantial act. Once this is shown and admitted, we have the substance of the whole of hylemorphism.

That there are in fact irreducible properties is acknowledged not only by the Scholastics but also by renowned scientists.[19]

3rd Argument. From the phenomena of crystallization. In crystals there is a certain power in virtue of which molecules are mutually disposed in a wondrous order and are ordered in a certain and determinate way; this is the power by which the crystal repairs and restores its angles when they are damaged. But this power, since it

[19] Cf. P. DUHEM, *Evolution de la mécanique*, 197-198.

dominates matter, is necessarily higher than matter. Therefore, above and beyond matter there is in crystals some formative power distinct from matter. Further, that power is not superior to the divine will: God, of course, rules His creatures in such a way that he puts in them the intrinsic principles of their actions; He is not their only cause, but beyond His will we must always assign a proximate and intrinsic cause. Therefore, in crystals there is some intrinsic power which we call a formal principle: "Thus crystallography supports the philosophical opinion expressed in the Thirteenth Century by the powerful genius of St. Thomas Aquinas."[20]

V. – COROLLARIES. From the preceding arguments it follows that there is not only a material principle and a formal principle, but, additionally, that nature is as the Scholastics describe it, namely, that matter is a substantial principle that is incomplete and that has the role of potency and that form is a substantial principle that has the role of act. It is manifest that each is a substantial principle. For out of their union arises a substantial composite. But it cannot be called a substantial composite if one of its components is an accident. *Therefore.* – Now, we conclude for the same reason that each is an incomplete principle. Something *per se* one results from the union of two things. But out of the union of two complete beings only something that is *per accidens* one can result, and not something that is *per se* one. Therefore, each is imperfect. But out of the two things a perfect being results. Therefore, it is necessary that one be conceived as perfectible, and the other as perfective, one as a rudiment (*incohatio*), the other as a complement. But those things that are compared to each other as something perfectible and something perfective, as rudiment (*incohatio*) and complement are related to each other as potency and act. Therefore, matter and form are related to each other as potency and act.

VI. – SUBSTANTIAL CHANGE. We demonstrated the three assertions of the Scholastics abstracting from substantial change. We do not stop here, but rather it is necessary to explain in detail the more intimate aspects of the matter from the fact of substantial change. Change in general is a transition from one mode of *esse* to another. A

---

[20] DE LAPPARENT, *Cours de minérologie*, p. 68: "Ainsi la cristallographie donnerait raison à l'opinion philosophique exprimée dès le treizième siècle par le puissant génie de S. Thomas d'Aquin."

transition to an accidental mode of *esse* is called *alteration*; but a transition from one nature to another is a *substantial change*, of which two kinds are distinguished: the change by means of which a subject obtains a substantial form is called *generation,* and the change whereby the subject loses a substantial form is *corruption.* Now, there is an axiom: "The corruption of one thing is the generation of another" (*Corruption unius est generatio alterius*), e.g., the corruption of a living thing is the generation of a carcass, and the corruption of a tree is the generation of ashes. The reason for this is evident. The subject cannot remain without form; otherwise, it would be altogether indeterminate. It is necessary, therefore, that the expulsion of one form be the introduction of another. Another unshaken axiom among Scholastics is: "Corruptible and material beings arise through generation, not creation; they perish through corruption, not annihilation" (*Entia corporalia et materialia generatione, non creatione, oriri; corruptione, non annihilatione, interire.*) Ash does not come to be out of nothing, but is generated from wood; the wood does not lapse into nothingness, but is corrupted. To these corresponds the dictum of the moderns: "Nothing is created; nothing is lost" (*Rien ne se crée; rien ne se perd*).[21]

VII. – ARGUMENT FROM SUBSTANTIAL CHANGE. Presupposing these things, one may argue thus: If there is substantial change, then prime matter and substantial form must be admitted. But there is, in fact, substantial change. *Therefore.*

Explanation of the Major. Since all change requires a subject that is otherwise after than before, if there is a substantial change, there must be some substantial subject that loses substantial *esse* through corruption and that acquires a new substantial *esse* through generation. But a subject that can lose and gain *esse* is in potency to it. Therefore, one must admit that there is some substantial principle that is ordered to another as potency to act, and we call that subject 'prime matter'. Secondly, there must be another substantial principle that loses through

---

[21] Cf. Treatise I, q. 2, a. 2, n. 11; q. 3, a. 3, n. 1; Treatise II, q. 1, a. 1, n. 5. P. DUHEM, *Le Mixte*, p. 205: "Scholasticism said, *Corruptio unius generatio alterius*; modern chemistry completes that same principle and makes it more precise by demonstrating that the destroyed mass is always equal to the created mass." (*Corruptio unius generatio alterius*, disait la Scolastique; la Chimie moderne complète et précise ce principe en nous montrant que la masse détruite est toujours égale à la masse créée).

corruption, or gains through generation, whatever completes the potential subject. But that which completes the potential subject is related to it as act to potency. Therefore, another substantial principle must be admitted that is related to matter as act to potency, and we call this 'form'. Substantial change, therefore, requires matter and form.

[Proof of the Minor.] That there is substantial change is proven in many ways.

1st Argument. There is substantial change when a new composite receives specifically diverse properties distinct from the properties of the previous composite. But there are many sets of successive beings (*adjuncta*)[22] in which the properties of the new composite are specifically distinct from the properties of the previous composite. Therefore, in many successive beings (*adjuncta*) there is substantial change. ***The reason for the major premise*** is that properties are derived from the substance. Therefore, where there are specifically diverse properties, there is a specifically diverse substance. That suffices for the *ratio* of substantial change. For one must not confuse substantial change with *transubstantiation*, in which the whole substance, both according to form and according to matter, is converted to another; but simple substantial change does not require that the whole composite that is in reality change, but only that the specific principle change, or that there be specifically diverse properties, which argue for a specifically diverse substance. ***Proof of the Minor.*** (1) In the case of living things, without a doubt there appear specifically diverse properties in generation and in corruption. The plant becomes ashes, the animal becomes a corpse; the properties of ash retain nothing in common with the specific properties of the plant, and the corpse possesses specifically diverse properties from the properties of the animal. From food, flesh is generated: flesh, of course, differs specifically from food. And it cannot be said that flesh is the juxtaposition of the parts of the food, for living things are nourished and grow, not through juxtaposition, but through assimilation. The mineral does not have in itself the principle of its evolution or of its motion, but is made and moved by another; but the living body evolves itself, and constitutes its own organism, such that life can be said to be a certain creation. "If it were necessary to define life with one word... I would say: life is creation... the creation of a machine that develops itself

[22] By *adjuncta*, Hugon appears to mean two beings that exist in succession; that is to say, one comes to be right after the other ceases to be. – ***The Translator.***

before our eyes..." (Claude Bernard). Even though the word 'creation' must be taken in the loose sense, nonetheless, the properties of the living thing are, without a doubt, essentially distinct from the properties of the mineral. Therefore, it is clearly established that every time that an *ens* passes from the mineral state to the living state, or vice versa, there are specifically diverse properties, and so there is substantial change. (2) Now, in the type of change that we observe in inorganic bodies it is not so evident that there are specifically diverse properties involved. Many neo-Scholastics concede that the issue is only 'probable' (*probabilis*), or 'more probable' (*probabilior*).

Now, it seems to us that the facts of chemistry argue for specifically diverse properties. This is argued thus.[23] *Constant, invariable, permanent*, and *irreducible* properties require a principle that is constant, invariable, permanent. But a principle that is *constant, invariable, permanent*, is nature, or *species*. Therefore, constant and invariable properties demonstrate the existence of a species of body. Therefore, if in one body are found invariable and permanent properties that in no way appear in another body, it must be concluded that in that one body there is a species that is not present the other, or that the properties the bodies are specifically diverse. Further, it often happens that in a new composite there are constant, invariable properties that in no way are found in the previous composite, or in the elements that compose it. Therefore, in these successive beings (*adjuncta*) there are specifically diverse properties. *Proof of the last minor premise.* Chemical compounds exhibit constant and invariable properties that in no way—indeed, not even with the aid of a microscope—can be discerned in the elements that compose it; and vice versa, the constant and permanent properties of the elements are no longer found in the compound. "Thus one can say that there is no oxygen [O_2], no sulfur [S_4], no phosphorus [P_4], no arsenic [As_4]—at least in the way we know them—in sulfurous acid [H_2SO_3], phosphine [PH_3], or arsine [AsH_3]."[24]

[23] The disagreement among philosophers comes from the fact that they do not agree on the notions of species and of specific properties. For this reason, in our proof we begin from the very notion of specific property, which must be *constant, invariable, permanent, irreducible*.

[24] SAINTE-CLAIRE DEVILLE, quoted from MIELLE in his extraordinary work, *De substantiae corporalis vi et ratione*, p. 170, note 1. Chemical formulae have been added for clarity. – ***The Translator.***

Similarly, water [H_2O] has fixed and permanent properties that are not found in oxygen [O_2] or hydrogen [H_2].[25]

The argument proposed above is worth repeating: In one body there is an interior order, tendency, finality, etc., that is distinctive in the new body into which the previous body changed: the finality found in the previous body is different from that found in the new composite; the old and the new composites are governed differently by inner their tendencies. But order and finality argue for constant and permanent, and hence specific, properties. Therefore, it is evident that in this kind of change there are specifically distinct properties involved.

Again, science finds in bodies certain *primary* and *irreducible* properties,[26] which are, thus, to be considered to be specifically diverse.

Proof. 1st Argument. Those things are diverse in species which require material and efficient causes that are essentially distinct, for a specific diversity in the effect is known through its distinctive material cause and through the diverse mode of acting of its efficient cause. But a chemical compound requires material and efficient causes of different order from the material and efficient causes that bring about a simple mixture. Therefore, the properties of the compound are specifically diverse from the properties of the simple mixture. *Proof of the Minor.* A mixture can come about in any quantity and proportion on the part of the material cause, but the compound requires certain and determinate proportions. To produce or separate mixtures, an agent that attracts or moves merely locally is sufficient,[27] but to produce compounds a more vehement agent is required that, through a very efficacious action, can intimately and profoundly affect the elements; and similarly to separate a compound an agent of superior power is required that is called a 'chemical reagent' [or 'reactant'].

2nd Argument. The perfection and beauty of the Universe, which consists in variety, requires that there be in things all degrees of beauty. But the degrees of beauty are such that, first, in the highest summit of things there must be an *ens* that is intrinsically immutable, in which neither accidental nor substantial can be conceived; second, such that there must be beings that are immutable as far as their substance, but mutable as far as their accidents; and third, such that there are mutable

---

[25] See below for a discussion on the permancence of elements.

[26] Scientists acknowledge this, for example, DUHEM, *Evolution de la mécanique*.

[27] LORENZELLI, *Philos. Nat.* P. I, lect. 4. n. 2.

beings both as far as their substance and as far as their accidents. Therefore, the beauty of the Universe requires that there be substantial change in the world.

And this beauty cannot be sufficiently replicated if substantial change happened only among living things. For it is inadmissible for inorganic bodies to be constituted in a degree of *ens* that is above that of living things. But inorganic bodies would be in a superior degree of being if living things were corruptible, but inorganic bodies remained immutable as far as their substance. Therefore, the order of the Universe requires that there be corruptions and substantial changes, not only among living things, but also and especially among inorganic bodies.

3rd Argument. It pertains to the dignity of secondary causes that one substance can produce another; of course, the causality of creatures would be imperfect if a substance could only produce accidents. But, unless there is substantial change, one substance would be unable to produce another. Therefore, the dignity of secondary causes requires that there be in things substantial change. *Proof of the Minor.* A creature does not produce something by creating, but by changing. Therefore, for it to produce an accident, an accidental change must occur; and for it to produce a substance, a substantial change must occur.

VIII. – CONFIRMATION. It is abundantly clear that the principles of bodies are prime matter and substantial form. It is further confirmed by the fact that in matter and form all the properties of first principles can be verified. *They are not from others*, for before matter and form nothing can be conceived in the natural body. – *They are not from each other*, for although matter and form are mutual causes, one does not enter into the essence of the other; but are mutually related as potency to act, which are opposites. – They are *first contraries*, both because they are substantial principles, and because they are opposed privatively, as we shall show in the last conclusion that is soon going to be presented. – They aptly *explain the apparent antinomies to which bodies are subject*. The fact that bodies are passive and inert is rightly explained through prime matter, which has the *ratio* of potency; the fact that they are active can be perfectly explained through form, which is actuality and the principle of act. The fact that they are divisible comes from the matter, which is the principle of diffusion; and the form, which is a simple power, is responsible for the fact that they possess unity. The fact that in them there is something common and generic is due to the matter, which is, like the genus, something potential and determinable; but the fact that they have specific properties results from form, which is like the difference and gives the thing its species. The fact that in them something always remains and something else passes is not difficult to conceive, in our opinion: the form passes, and there remains the matter, which is presupposed in every generation and which is left behind in every corruption. The fact that, after a chemical reaction occurs, the same weight remains, can be explained in the same way: weight and quantity are derived from matter; therefore, since matter always endures, those things must also remain of which it is the principle and root. On that account, the Laws of Chemical Combinations [namely, the Law of Conservation of Mass, the Law of Definite Proportions, the Law of Multiple Proportions] and the Law of Equivalent Weights, are indeed saved in the scholastic system.

Recent scientific experiments regarding what they call 'ions' seem to be favorable to the scholastic system to a certain degree. Indeed, the issue is not devoid of obscurity, and it is not possible yet to derive an altogether certain conclusion; but yet from these facts one can probably (*probabiliter*) derive the conclusion that in bodies two principle coincide: namely, *potentiality* in matter (*ions*) and *actuality* in matter (*electron*). Further, the illation from potency to prime matter, and from actuality to form, is legitimate.

IX. – Last Conclusion: "Beyond matter and form, which are positive principles, a certain negative third principle is rightly established, which is called 'privation'."

Proof. Since generation is a transition from substantial non-being to substantial being, three principles are required: the transient subject, the substantial non-being as the terminus that is left behind, and the substantial being as the terminus that is acquired. But the transient subject is *matter*; the terminus that is left behind is the lack of the form that is arriving, and this lack of form is a *privation*; and, finally, the terminus that is acquired is the arriving *form*. Therefore, there are three principles: *privation*, *matter* and *form*. Privation, however, is a principle in becoming, and is not in *esse* in fact; for once the form arrives, the privation is excluded.

X. – Difficulties Resolved; Whether there is a Contradiction Between the Scholastics and the Scientists.

1st ***Objection***. From the phenomena of allotropy and isomerism it would seem that the changes that occur in bodies do not involve a diverse nature, but only a diverse state in the same nature. But where a diverse nature is not involved there is no substantial change. Therefore, the aforesaid phenomena show that the changes of bodies are not substantial. *Proof of the First Part of the Major. Allotropy* is a phenomenon by virtue of which a body receives new properties and, nonetheless, remains the same substance. There are two classic examples: When oxygen [O_2] becomes ozone [O_3], it receives new properties and nonetheless retains the same nature; when *amorphous phosphorus* changes into *red* phosphorus, it acquires new chemical properties and nonetheless it remains in the same species of phosphorus. *Proof of the Second Part. Isomerism* is a phenomenon by virtue of which different bodies, though having different properties, are, nonetheless, made out of the same elements and in the same proportion: *cellulose, dextrin, starch, gum acacia,* are made out of the same elements in almost the same proportions. Therefore, the fact that there are different properties does not imply the presence of a new species of substance, but of a different state or mode.

Reply. *I distinguish the major.* (1) That the aforesaid phenomena show that *some* changes argue for a new state, let it be so [for the sake of argument]. But that *all* changes argue only for a new state, I deny. *I concede the minor premise. I distinguish the conclusion.* That the aforesaid phenomena prove that *some* of the changes in bodies are not substantial, let it be so [for the sake of argument]. But that *none* of the changes in bodies are substantial, I deny.

However evidently Chemistry shows that there is no substantial change in *allotropy* and *isomerism*, from that nothing can be inferred against the scholastic system. For we do not argue that all the changes in bodies are substantial; we, of course, admit that in many events it is

---

[28] "Allotropes" are different structural arrangement among atoms of the same chemical *element*; e.g., graphite and diamond are allotropes, or different structural arrangements, of carbon (C). In contrast, "isomers" are different structural arrangements among molecules of the same chemical compound; e.g., butane and isobutane are isomers, or different structural arrangements, of the hydrocarbon C_4H_{10}. The basic philosophical point is this: when two or more chemical substances are allotropes or isomers, they have the same constituents but different structures, and thus, they usually have very different properties. Hence, the philosophical question arises whether they belong to different species. – ***The Translator.***

difficult to discern whether there is substantial change or not; but it does not pertain to us to determine what occurs in each case. But we hold that many successive beings (*adjuncta*) involve specifically diverse properties, and thus substantial change; and this we have established this through many arguments.

This observation would be sufficient to resolve the difficulty.

(2) But what the adversaries assert, namely, that the aforesaid phenomena argue only for a new state, and not a new nature, can be denied. Oxygen and ozone, amorphous phosphorus and red phosphorus, materially remain the same substance; but they are formally and specifically diverse. For indeed, the ceasing to be of the preceding properties and the production of new, constant, and permanent properties argue for a new species. But in allotropy, the preceding properties cease to be and new properties are produced. Hence it is that Liebig, the chemist, said that oxygen [O_2] differs as much from ozone [O_3] as from chlorine [Cl_2].

Further, the phenomena of isomerism show that the same elements are in potency to become diverse bodies, but they do not prove with certainty that the aforesaid elements remain in act according to their proper forms, as we shall later say of the permanence of elements in a mixture. Moreover, it is impossible that diverse bodies be produced solely out of a diverse placement of the same elements, for placement is an accident and thus requires a body that is already constituted. Therefore, in isomers is required some formal and specifying principle from which new and diverse properties are derived.

2nd Objection. The principles of bodies must be made manifest through experience. But matter and form are not made manifest through experience. Therefore, they are not the principles of bodies. ***Reply***. *I distinguish the major.* That they must be made manifest through experience with reason as an intermediary, I concede; but that they must be made manifest through experience alone, I deny. *I contradistinguish the minor.* [That matter and form are not made manifest through experience alone, I concede; but that they are made manifest through experience with reason as an intermediary, I deny.] *And I deny the conclusion.* – In this the adversaries are deceived because they think that the question of principles is a physical and experimental one, whereas it is really a properly philosophical one. For indeed, the resolution of the problem of essential principles pertains to that science to which it is proper to consider essence, or quiddity. But essence is not subject to experience, but only to intellect, and, consequently, its consideration is to be ascribed to philosophy. It is not required, therefore, that the principles of bodies be known through experience alone. Because our knowledge arises from sensibles, the principles of bodies must be made manifest through experience with reason as an intermediary. Moreover, the Scholastics also use experience, insofar as they conclude, from the properties that experience finds in bodies, that there are a material principle and a formal principle.

3rd Objection. Apart from the Scholastics who are ignorant of natural things, no one else professes hylemorphism. Therefore, it is prudent to mistrust this system. ***Reply***. In this question we must believe the philosophers more so than the physicists and the chemists, as is evident from the previous reply. Further, the greatest philosophers, Aristotle, St. Augustine, and St. Thomas, adhered to this system. Even today many outside of of the Scholastics support it. Barthélemy Saint-Hilaire presents this testimony of the Aristotelian doctrine: "For me, I find it to be simple and true, and it does not have the fault of being obscure; at most, I will grant that it has a certain subtlety, without being in any way sophistical. Matter and form are the logical and real elements of being."[29]

Moreover, not all scientists are opposed to hylemorphism: on the contrary, some of the most learned men in science are favorable to it, e.g., A. DeLapparent, P. Duhem, etc. But it often happens that scientists attack the scholastic system because they are too unfamiliar with the scholastic disciplines, especially Metaphysics. But scholastic philosophers, especially the more recent, are not that unfamiliar with the natural sciences, and generally are more apt to make judgments concerning the philosophical opinions of the scientists than the scientists themselves are able to decide on the metaphysical principles of the Scholastics.[30]

Still, the contradiction that many imagine exists between scientists and Scholastics can be denied, because scientific atomism does not oppose hylemorphism, but treats quite a different object. Scientists, of course, who aim at proximate causes only, need not trouble about matter and form, which are in the order of first principles. Therefore, since scientific investigation finds only atoms in bodies, it can be rightly

---

[29] *Préface de la Physique*, p. 28.

[30] For other difficulties, see NYS, *Cosmol.*, L. II, ch. 5; *Revue néo-scolastique*, 1904.

established that atoms are the principles of bodies, that is, proximate principles, for in science one does not trouble with other principles.

The Scholastics, however, speculate about a higher object, the quiddity of things, which is above the order of physics and chemistry; hence, conceding that atoms are the proximate principles of bodies, they further inquire concerning the prior principles, which are discovered, not by experience, but by philosophical and metaphysical processes, and which reason shows they are related as substantial potency and substantial act, or as matter and form.

For this reason, if scientists and Scholastics stay in their respective order, namely, scientists in the order of proximate principles, which are made manifest through experience, and Scholastics in the order of essential principles, which are the objects of philosophy, a contradiction will never occur, nor will there be disagreements; but the controversy will have to be stopped by the philosophers themselves. Meanwhile, let it be clear that scientists that are true to their name do not contradict hylemorphism: "Current physics tends to recover a certain form of peripateticism" [i.e., Aristotelianism].[31]

----

[31] P. DUHEM, *Le mixte*, p. 200.

SECOND QUESTION

On the constitutive principles of bodies in particular

Now it is necessary to examine more intimately and fully the notions of matter, of form, of the composite that results out of them, and of substantial generation, which is so important in the scholastic system.

FIRST ARTICLE

On prime matter

I. – The Analogical Concept of Prime Matter. Just as we know invisible things through an analogy with visible things, so prime matter will be known through an analogy with secondary matter. Further, secondary matter is a composite, or a sensible body, which is endowed with power and extension. Now, we see that diverse determinations are received in a body, e.g., that water transitions from coldness to a state of heat, and that marble obtains different configurations. This forces upon us the concept of some subject that transitions from one state to the other, and so, which is in potency to that to which it proceeds. But secondary matter is the subject of accidental change, or of the transition from accidental non-being to accidental being. Only through an analogy, therefore, the mind imagines for itself a subject of substantial change, in which occurs a transition from substantial non-being to substantial being. Moreover, the subject of substantial change must be substantial and, because it can lose or gain being, it is in potency to it. Therefore, thus is formed in the mind the concept of some substantial subject that is related to the substance in the manner of potency. The Philosopher uses this process of reasoning to define prime matter in *Physics* I.

II. – The Definition of Matter. For that reason, matter is, according to Aristotle, "the first subject of each thing, out of which, when it is in it, something comes to be and not *per accidens*" (*subjectum primum cujusque rei, ex quo, cum insit, fit aliquid et non per accidens*).

It is said to be: (1) a *subject*, which is the genus of the definition, and it coincides with the matter; (2) the *first* subject, whereby it is distinguished from secondary matter: for the secondary matter is a subject of accidents, whereas the first subject is not, because it receives substantial form; (3) *when it is in it*, so that it may be distinguished from privation and hence so that it be also indicated that matter is not a negative terminus, but an *intrinsic* principle, entering into the composition of the thing; (4) *out of which something comes to be and not* per accidens; this expression means that the composite that is produced out of matter and form is not one *secundum quid*, but is *per se* and *simpliciter* one, that is to say, something that expresses one complete essence.

There is another negative definition of matter: Prime matter in itself "does not have a 'whatness', or quality, or quantity, or any of those things whereby a being is determined" (*non est quid, nec quale, nec quantum, nec eorum quibus ens determinatur*). It does not have 'whatness' (*non est quid*, literally, "it is not a 'what'"), that is to say, it is not a quiddity or a substance with a complete species: it is not an immaterial substance, which does not undergo change from one substantial being into another; nor is it a material substance, because a substance is precisely composed of matter and form; nor is it secondary matter, which is the subject of accidents. Prime matter is the subject of substantial change. – It does not have quantity (*non est quantum*, lit., "it is not a 'how much'"), because quantity is an accident which requires a corporeal substance already constituted in its *esse*. – Nor does it have quality (*nec est quale*, roughly, "it is not a 'what kind'"), because a quality requires a substance as well as a quantity in which it is immediately received. – Nor is it any of those things whereby an *ens* is determined, that is to say, it is not any of the categories of accidents that affect and modify *ens*, such as relation, action, passion, etc.; these are indeed extrinsic and transient, but matter is a intrinsic and permanent principle. Hence, according to itself, prime matter can be directly placed in any of the predicaments. It belongs, however, reductively to the predicament of substance. For it is an intrinsic part of corporeal substance; and parts are referred reductively, to the predicament of the whole.[32]

---

[32] Cf. *Metaphysics* 8.3; ST. THOMAS, *In VIII Metaph.*, lect. 3.

III. – WHETHER MATTER IS POTENCY. All Scholastics acknowledge that matter is a potency that is receptive of form, but is it pure potency? We must bear in mind that pure potency can be understood in two senses: first, as designating a merely logical potency, that is, a mere non-impossibility of existing; in this sense it is clear that prime matter is not a pure potency, because it is a certain substantial reality. Second, as meaning a real potency, but only a passive one, which of itself does not include act. And act is twofold: namely, that of essence and that of existence. The act of essence, or entitative act, is that which constitutes the thing in its quiddity, in a certain and determinate mode of *esse*; and the act of existence places the thing outside of the state of possibility and constitutes it in the *ratio* of actuality in reality.

Scotus, Suárez, and many others thought that to matter belongs a certain partial existence and a certain incomplete entitative act. But St. Thomas, the Thomists, and the more recent Scholastics unanimously deny this.

IV. FIRST CONCLUSION: "Prime matter does not have its own existence."

Proof from St. Thomas: "To say that matter is in act without form is to say contradictory things simultaneously."[33]

Proof from Reason: A substantial composite that is *per se* one results from the combination of matter and form. But if matter has its own existence, then it is neither a substantial composite nor a composite that is *per se* one. *Therefore. Proof of the First Part of the Minor Premise.* If matter has its own *esse*, then the form gives only an adventitious, adjacent, and secondary *esse*. But that which gives an adventitious, adjacent *esse*, is not *esse simpliciter*, but only accidental *esse* or *esse secundum quid*. Therefore, the form would only give to matter an accidental *esse*, and so substantial *esse* could not result from the combination of matter and form. *Proof of the Second Part of the Minor Premise*. That is *per se* one which has a single substantial *esse*. But if matter enjoys its own being, then a double composite would receive substantial being, namely the substantial *esse* of matter and the substantial *esse* of the form. Therefore, it would not be *per se* one.

[33] *Quodlibet* 3, q. 1. a. 1, at the end of the body of the article: "Dicere ergo quod materia sit in actu sine forma, est dicere contradictoria esse simul...."

And one cannot reply that the existence of matter is incomplete and merely a means to obtaining form (*incompleta et vialis*), and that for that reason it is possible for the composite of matter and form remains *per se* one. For, indeed, existence is the last act and the last terminus of an *ens*. But the last terminus of an *ens* belongs only to a thing that is substantially complete and that has reached its terminus (*completa et terminata*). Therefore, existence belongs only to a thing that is substantially complete and that has reached its terminus (*completa et terminata*). Therefore, if matter had its own existence, it would have reached its terminus substantially and hence it would not be able to constitute, together with the form, something that is *per se* one.

Confirmation. Existence presupposes the act of essence (*actus essentiae*). But matter does not have the act of essence (*actus essentiae*) independently from form. Therefore, neither does it have the act of existence (*actus existentiae*). The minor will be established from what we shall say, whence is the second conclusion.

V. – SECOND CONCLUSION: "Prime matter does not have its own entitative act."

Proof from St. Thomas. "It is clear that matter alone is not an essence."[34] — "Matter does not properly have an essence."[35]

Proof from Reason. Every entitative act is either subsistent (that is, spiritual), or informing (*informans*), or accidental, or substantial. But matter is not a subsistent act, as is evident; it is not an informing act (*actus informans*), which, of course, would have to be received in some matter, which itself would again require an entitative act that would have to be received in another matter, and thus the process would go on *ad infinitum*. Similarly, matter is not an accidental act, since it is a substantial principle; nor is it a substantial act, because the first substantial act is the form, and the second substantial act is existence. And matter is neither form nor existence. – Further, substantial act is act *simpliciter,* because it is the basis of all things and constitutes a thing in its first and substantial *esse*. Therefore, if matter were a substantial act, it would be an *ens simpliciter*, and it would not make, together with the form, something that is *per se* one. It remains, therefore, that matter is

---

[34] *De ente et essentia* 2: "Quod materia sola non sit essentia, planum est."

[35] *De veritate*, q. 3, a. 5: "Materiam proprie non habere essentiam."

only a substantial potency, or, as has been said many times, an incomplete substantial principle that has the role of passive potency.

VI. Objections. ***1st Objection***. Matter is either something (*aliquid*) or pure nothing. If it is nothing, then it is not to be considered to be the first principle of bodies; if it is something, then it has some essence. Therefore, matter exhibits some entitative act. ***Reply***. That if it is something it must have some determinate essence *simpliciter*, that is, a specific mode of *esse*, I deny. That if it is something it must have some essence whose entire *ratio* consists in the fact that it is pure potency, I concede. That therefore it is an act, I deny.

Matter is certainly something real, but only in the order of potentiality; that is to say, it is a real potency, which nonetheless cannot exist or be understood without form, just as potency cannot exist or be understood without act.

2nd Objection. But all reality can be said to be an act. Therefore, matter is an act. ***Reply***. No. A reality that is of itself indifferent to all modes of *esse* cannot in any way be called act, but only that which is determined or determining; otherwise, there would be no difference between real potency and act.

3rd Objection. Matter differs from form. But a difference is an act. Therefore, matter has act. ***Reply***. *I distinguish the major.* That matter differs from form through a [specific] difference, strictly speaking, I deny. That they differ through a difference in a broad sense, I concede. *I contradistinguish the minor.* [That a specific difference, strictly speaking, is an act, I concede. But that a difference, in the broad sense, is an act, I deny.] *And I deny the conclusion.* Matter, properly speaking, does not have a genus or a [specific] difference, but differs from form as potency differs from act. But, potency is not said to be act, due to the fact that it differs from act. Therefore, matter cannot be said to be act because it is distinguished from form.

VII. – Two Corollaries. ***1st Corollary***: *Matter cannot be known in itself.* For each thing is known insofar as it is in act, because intelligibility is founded on the *ens*; but *ens* is said of the thing that is in act. Therefore, since matter is not of itself in act, it can only be understood through its relationship to substantial form, by whose *ratio* it is determined to some *esse*.

2nd ***Corollary***: *Even for God's absolute power, matter cannot exist without some form.* This is the opinion of the Thomists, which is opposed to that of the Scotists and many others. Thus says the Angelic Doctor: "God cannot cause something that both is and is not, or something involving a contradiction, and *such is the case of matter existing without form.*"[36]

Even for God's absolute power, it cannot happen that something exists without any act of essence or existence (*actus essentiae et existentiae*). But it has been established that matter without form cannot possess any act of essence or existence. Therefore, it is contradictory for matter to exist without any form.

Further, it is absolutely contradictory that there be an *ens* that is not constituted in any specific degree of *ens*; for there are no general realities, but only concrete realities that have determinate species. But matter of itself is not in any specific degree of *ens*, since all specific degrees are established by the form. Therefore, it is absolutely contradictory for matter to exist without form.

Objection. An accident can exist without a subject by divine intervention. But matter is less dependent on the form for existing than an accident on its subject. Therefore, matter can exist without form by divine intervention. ***Reply***. *I deny the minor.* For an accident is a certain act, even if imperfect, whereas matter is pure potency. Now, an imperfect act is less dependent on a perfect act than a pure potency depends on the first act. The reason is that an act, even if imperfect, has some essence and a determinate degree of *ens*; but pure potency is indifferent to all modes of *esse*.

VIII. – The Attributes of Matter. Matter has transcendental properties, though not completely and perfectly except through its relation to form. For perfect properties are convertible with being. But matter is properly an *ens* through form. Therefore, it obtains unity, truth (or intelligibility), and goodness through its relation to form. Matter is said to be *one* and specifically common insofar as it is of the same *ratio* in all bodies and can be present in all degrees of beings, "as that which was in the form of water and earth later becomes bitumen, then fire, and

---

[36] *Quodlibet* 3, q. 1, a. 1: "[Q]uod aliquid simul sit et non sit, a Deo fieri non potest, nec aliquid contradictionem includens. *Et de huiusmodi est materiam esse in actu sine forma.*"

afterwards air."[37] This is a *negative* unity, however; that is to say, matter is one is species, not because it has a principle of unity, but because it lacks all act that would distinguish it. It can also be said to be one in number negatively, because it is understood without all those dispositions that make a thing differ in number.[38]

The other attributes of matter are these:

(1) *It cannot be generated or corrupted.* When bodies are generated, they are not made out of nothing, and when they are corrupted they do not fall into nothingness. Therefore, it is necessary that every generation presupposes matter and that every corruption leaves behind some matter. Therefore, matter can neither be generated nor corrupted; but because it always desires a new form, it is said to be the principle of corruption.

(2) *It is simple as far as its essence*, that is to say, it cannot be analyzed into many elements out of which it is constituted, otherwise it would not be a principle, but a *principiatum*.

(3) But yet it is the *root of quantity*, not because it has quantity in act, but because it is the substantial principle from which extension emanates as a property. For by the very fact that form is united to matter, a substance that is extended into parts comes to be; indeed, matter depends first on this quantity than on this form.[39] Thus the concept of matter and the concept of extension are naturally associated. Hence, the dictum of the Angelic Doctor: "Quantity is received from matter and quality from the form."[40]

(4) *Matter is the principle of passivity.* An *ens* is passive insofar as it is in potency. But matter is in potency. Therefore, it is the principle of passivity; hence the axiom: "Matter is of itself inactive; it is for being, not for acting" (*Materia est de se otiosa, est ad essendum, non ad operandum*).

---

[37] GOUDIN, *Phys.* q. 3, a. 5: "[U]t quae erat sub aqua et sub terra, postea tamen unitatis principium, sed quod creat omni actu distinguiente. Dici potest etiam una numero negative, quia intelligitur sine omnibus dispositionibus quae faciunt rem differre numero."

[38] Cf. ST. THOMAS, *De principiis naturae*, c. 1.

[39] See below, q. 4, a. 1, n. 11. We shall discuss below, in q. 3, a. 1, whether a material substance has integral parts independently from quantity.

[40] *In IV Sent.* d. 12, q. 1, a. 2, qc., 1c: "[Q]uantitas se tenet ex parte materiae, et qualitas ex parte formae." – ***The Translator.***

(5) *Matter has an innate appetite for form.* Appetite is twofold: *elicited* and *innate.* An *elicited* appetite is an act of an appetitive potency, whether of a sensitive appetite or of a rational appetite, whereby something is carried over to some known good. An *innate* appetite is not properly an act, but rather a certain inclination or tendency instilled by the Author of nature into each thing, towards its own convenient good. In this latter manner matter tends to (*appetit*) its own perfection. But matter obtains its perfection only from the form. Therefore, it tends to (*appetit*) the form through this innate appetite; and, because no form is so perfect that it can satiate and put to rest the capacity of matter, matter thus always tends to an ulterior form, although it does not seek all forms in the same way. It tends to (*appetit*) the forms that it does not yet have with a quasi-desire. For desire is a tendency towards an absent good that is possible of attainment. But forms that are not yet had are, with respect to matter, an absent good that is possible of attainment, because they can be introduced through the action of the agent. – But matter tends to (*appetit*) the forms that it has in act in the manner of possession and rest; but it does not remain fully satiated and therefore there always remains an inclination to other forms.

Finally, matter retains a certain desire with respect to the forms that it had and has lost, for there always remains a proportion between that matter and those forms; but it is an inefficacious desire because there is no potency for the past, nor is there any natural agent that can reproduce numerically the same form.

But the 'formal *ratio* under which' (*ratio formalis sub qua*) matter seeks all forms is generable and corruptible substantial *esse*, in which all material forms coincide. This is the unanimous opinion of the Thomists.

Note, however, that this appetite is merely passive, and matter can develop only through the action of the agent.

IX. – The Identity and Permanence of Matter.

Objection. Matter is thought by Scholastics as being the same in all beings, and nonetheless experience attests to the fact that each of them has its own matter; the Scholastics further teach the permanence of matter, while science proves that matter is perpetually subject to flux and that it continuously varies. Therefore, hylemorphism contradicts experience and science. ***Reply***. When we established that matter is the same in all things, we did not mean that the real and numerically one matter of one body is not distinct from the numerically one matter of another body; no, in fact we defend the specific (or formal) and numeric (or material) distinction of bodies. Rather, the discussion concerns generic identity and universality, or universality of similitude only, insofar as bodies certainly differ from each other, but they receive their *esse* and their specific degree [of being] only from the form. By reason of their matter they coincide in something common, and by reason of their form they are placed in their own, specific degree [of being]. Now, science does not contradict this; rather, science confirms it, because it proves that there is in bodies a permanent and common principle, as well as a specific principle.

Regarding that which pertains to the permanence of matter, P. De San observes that, "*Permanent* matter is prime matter, considered bare and abstractly. *Transient* matter is prime matter, insofar as, by virtue of the form under which it is found to be in act, it is rendered capable to receive within itself another form. And this matter is itself either *remote* or *proximate*. Thus, for example, the elementary bodies are the most remote matter of the generation of an animal; the more proximate matter is the semen, and the most proximate matter is the embryo that it is led to the last dispositions."[41]

X. – The Teaching of St. Augustine on Prime Matter. Augustine gave great thanks to God for the understanding concerning prime matter that was divinely granted to him. For, before, when he heard the term, he did not understand the concept, but now, relying on divine aid and observing the mutability of things and their transition from form to form, he understood that matter is their subject of change. He teaches four things concerning matter: (1) it is the subject of change; (2) it is potency that is destitute of all act: "something unformed," which is not an *ens*, properly speaking; (3) but it is not a merely logical potency, "not altogether nothing," but something real and physical, from which visible and composite forms are educed; (4) it is really distinct from form, for matter is something unformed, the subject of the transition from form to form and that from which forms are educed. It is worthwhile to quote his words:

---

[41] *Cosmol.* n. 103: "Materia *permanens* est materia prima nude et abstracte spectata. *Transiens* est materia prima, prout ratione formae sub qua actu invenitur, redditur remote vel proxime apta ad recipiendam in se formam aliam. Eaque ipsa est vel *remota* vel *proxima*. Sic, v.g. materia remotissima generationis animalis sunt corpora proxima est embryo ad ultimas dispositiones perductus."

"But were I, O Lord, by my mouth and by my pen to confess unto You the whole, whatever You have taught me concerning that matter, the name of which hearing beforehand, and not understanding (they who could not understand it telling me of it), I conceived it as having innumerable and varied forms. And therefore did I not conceive it; my mind revolved in disturbed order foul and horrible forms, but yet forms; and I called it formless, not that it lacked form, but because it had such as, did it appear, my mind would turn from, as unwonted and incongruous, and at which human weakness would be disturbed. But even that which I did conceive was formless, not by the privation of all form, but in comparison of more beautiful forms; and true reason persuaded me that I ought altogether to remove from it all remnants of any form whatever, if I wished to conceive matter wholly without form; and I could not. For sooner could I imagine that that which should be deprived of all form was not at all, than conceive anything between form and nothing—neither formed, nor nothing, formless, nearly nothing. And my mind hence ceased to question my spirit, filled (as it was) with the images of formed bodies, and changing and varying them according to its will; and I applied myself to the bodies themselves, and looked more deeply into their mutability, by which they cease to be what they had been, and begin to be what they were not; and this same transit from form unto form I have looked upon to be through some formless condition, not through a very nothing; but I desired to know, not to guess. And if my voice and my pen should confess the whole unto You, whatsoever knots You have untied for me concerning this question, who of my readers would endure to take in the whole? Nor yet, therefore, shall my heart cease to give You honor, and a song of praise, for those things which it is not able to express. For the mutability of mutable things is itself capable of all those forms into which mutable things are changed. And this mutability, what is it? Is it soul? Is it body? Is it the outer appearance of soul or body? Could it be said, Nothing were something, and That which is, is not,

I would say that this were it; and yet in some manner was it already, since it could receive these visible and compound shapes. And whence and in what manner was this, unless from You, from whom are all things, in so far as they are? But by how much the farther from You, so much the more unlike unto You; for it is not distance of place."[42]

----

[42] *Confessiones* 12.6-7, Translated by J.G. Pilkington, in *Nicene and Post-Nicene Fathers, First Series, Vol. 1*. Edited by Philip Schaff. (Buffalo, NY: Christian Literature Publishing Co., 1887). Cf. also *De Genesi ad litteram*, chs. 15 and 19.

SECOND ARTICLE

On substantial form

I. – The General Notion of Form. Taken most broadly, form means act or perfection. Now, act is that which gives *esse* to a thing, and is distinguished into many kinds: subsistent act, if it receives *esse* separately from all matter; informing act, if it is received in matter; first act, if it gives *esse simpliciter* to that which does not presuppose anything else; second act, if it gives *esse* to that which presupposes something prior. Form is also defined by Aristotle as "that which is a species and exemplar." "Species" here signifies an intrinsic form that constitutes a thing in a certain determinate species, and "exemplar" designates an extrinsic form, namely, that in imitation of which something is made.

II. – The Concept of Substantial Form can be Obtained by Way of Analogy and Negation. Just as we considered the substantial subject of change through an analogy with the subject of accidental changes, so substantial form can be known from accidental form. For even if form is of itself knowable, those things that we first apprehend are operations and properties, that is, accidental forms. Now, that which accidental form provides is the determination of the subject to some mode of *esse*, giving it some species, not *simpliciter*, but in the *ratio* of such and such a being: for example, figure gives to marble a species in the *ratio* of statue. Therefore, we infer that substantial form must limit an *ens* to a certain mode of *esse* and to a certain species; and, because it is substantial, it must limit it to a substantial mode of being and to a species, *simpliciter*. From this we are led to describe form as *a certain substantial reality that constitutes the thing in a certain and determinate mode of esse and in a certain and proper species.*

Substantial form is also known by way of negation, that is, through its opposition to accidental form. Now, there is a threefold difference—of great importance, indeed—between these two. First, substantial form makes *esse simpliciter*, whence its subject depends on it for its existence, and of itself it is only in potency; but the accidental form makes something *be such* or makes it be *so much* (*esse tale aut esse tantum*), and its subject is an *ens* in act, which does not depend on the accidental form for its existence. Second, substantial form is not for the sake of its subject, but, on the contrary, matter is for the sake of the form; but accidental form is for the sake of the completion of the subject. Third, the substantial form is not received in varying degrees (*magis et minus*); for Peter is not sometimes more a human being and sometimes less a human being; but the accidental form can become more or less intense.

III. – Definition of Form. According to the Philosopher, substantial form can be defined as "the first act that, together with prime matter, constitutes something that is *per se* one." The genus of the definition is *act*. The words "together with matter" are added so that it may be distinguished from subsistent act; through the words, "constitutes something that is *per se* one," it is distinguished from accidental act, that is, the act whereby something becomes one *per accidens*; and, finally, through the words *first act*, it is distinguished from the act of existence, which is the second act.

Many other definitions are offered. The proper function of the form is to constitute the thing in a determinate mode of *esse* and to give it its species; hence, the form itself is often called *species*. As the thing is constituted in its species, it receives a determinate quiddity; hence it happens that the form is said to be the *ratio of quiddity* (*ratio quidditatis*). And by the very fact that it constitutes an essence, it causes that such and such properties emanate out of the essence; hence, it is also called the *root of properties* (*radix proprietatum*).

IV. – *Esse* Belongs to Form *Per Se*. St. Thomas says this in many places.[43] This is to be understood in two senses: (1) *Esse* belongs to form, not through another (*non per aliud*), but immediately; (2) not *per accidens*, but necessarily. *Immediately*, that is to say: between the form and the existence of a thing there is no intermediary action that joins the two, because it is not the case that the form first comes to be and then through another action it is joined to existence; but rather, as soon as the form is produced, it is joined to *esse*. Matter and form, on the contrary, are joined through an intermediary action, for in that case the agent joins the two preexisting subjects by his action.[44] Therefore, *esse* belongs to form immediately, and not through another (*per aliud*). –

---

[43] See *Summa theologiae* Ia, q. 50, a. 5; Ia, q. 75, a. 6.

[44] Cf. Cajetan, *In Iam Partem*, q.50, a. 5.

Not per accidens. The sense is not that form exists by its own power (*vi sui*), as if existence were an essential predicate of the form, since to God alone belongs existing by His own power (*vi sui*); rather, the sense is: the form having been produced, *esse* is joined to it necessarily, and not *per accidens*. Form, of course, in its essential concept, is an act that gives *esse*. Therefore, supposing that it is already produced, it must necessarily be in act, and not have *esse* in a merely *per accidens* way. Therefore, *esse* belongs *per se* to form. But that which belongs to something *per se* is inseparable from it. From this the other property of form follows.

V. – *ESSE* IS INSEPARABLE FROM FORM. Corruption occurs, not because being is separated from form, but because the form is separated from the subject. Form and being go together, such that it is true to assert: If the subject loses its form, it loses its *esse*, and if it loses its *esse*, it loses its form; similarly, if it retains its form, it retains its *esse*, and if it retains its being, it at the same time retains its form. Therefore, since to acquire or to lose being is to acquire or to lose form, and vice versa, it is clearly established that *esse* and form are inseparable. We already have given a reason. *Esse* can certainly be separated from matter and from the composite, because it does not belong to them *per se*, but by means of the form; thus, matter and form can be separated from each other, because they are united by means of the action of the agent. "But wherever two things are related in such a way," says Cajetan, "that they are joined universally without some intermediary action that does the joining, then those two things are necessarily inseparable; therefore, they are both generated by the same generation and are also corrupted by the same corruption." Therefore, since *esse* and form are immediately joined, it is impossible to separate them; they will either perish together or will persevere together.

Therefore, if there were some form that existed without matter, it would not be subject to corruption. For *esse* cannot go away unless the form goes away. But form, which can certainly be separated from matter, nonetheless cannot be separated from itself. Therefore, corruption, which can certainly take place where form is in matter, is impossible wherever form subsists without any matter. For this reason, it is firmly proven that angels and human souls are incorruptible.

From the foregoing we can infer the nobility of form. Since *esse* belongs to form and a thing has a likeness to God insofar as it has *esse*, hence it is that form is said to be *something divine and desirable* (*divinum quoddam et appetibile*).

VI. – Activity Belongs to Form. *Esse* acts insofar as it is in act. But it is in act through form. Therefore, the composite does not act by reason of matter, but by reason of the form, which is act and the principle of act.[45]

Further, that which gives species also gives the actions that manifest the species. But form gives the species. Therefore, it is also the principle of operations. Hence the dictum: "Each thing is passive by reason of matter and acts by reason of the form" (*Unumquodque patitur ratione materiae, agit ratione formae*).

VII. – Simplicity Belongs to Form. Form is said to be simple in two senses: (1) because it does not consist of many entities into which it can be resolved; for one cannot conceive a form that is composed of matter and another form; (2) because it cannot be *per se* divided. Divisibility, of course, is due to matter, which is the root of quantity; and division itself is done through quantity. But form is that whereby an *ens* is constituted as something that is one (*aliquid unum*) and undivided in itself. Therefore, divisibility does not belong to form; rather, it is the principle of unity. Hence, if atoms cannot be divided, that is not due to matter, which is the principle of division, but is due to the form, which is the principle of indivisibility. But yet, the form is *per accidens* extended and is subject to a place by reason of the composite. Hence, it is not impossible for inferior forms, even though they are *per se* indivisible, to become *per accidens* divisible.[46]

VIII. – The *Ratio* of Form Precludes that there be Many Substantial Forms in the Same Composite. This is the teaching of St. Thomas and of many Scholastics. There are three opinions on this matter.

The first admits many forms in the composite that are essentially subordinate, the inferior to the superior. This is the view of Avicenna and Averroes. Close to this view is the opinion of Scotus, who posits in the organic body the form of 'corporeality' and the form of 'living'. The

[45] Cf. St. Thomas, *In I Sent.*, d. 12, q. 1, a. 2.

[46] We shall discuss whether the forms of living beings are divisible in IIa-IIae, Treatise 1, q. 3, a. 3.

second opinion claims that those multiple forms are not essentially subordinate but are related to the principal form as its dispositions. Thus, in man, according to many moderns, there is a rational soul as the primary form and additionally the forms of chemical elements as the soul's dispositions. The third opinion upholds one substantial form in one matter.

We prove St. Thomas' view from his very words: "To one thing belongs one substantial *esse*. But the substantial form gives substantial *esse*. Therefore, to one thing belongs only one substantial form."[47] Further, as the Scholastics unanimously argue, one and the same substance cannot be essentially constituted in two ultimate species. But a substance is constituted in its ultimate and complete species through any one form. Therefore, there cannot be many essentially subordinate forms in one matter.

Nor can one be a disposition to another. For the form that constitutes the complete thing *simpliciter*, in the *ratio* of substance, is not a disposition to another. But any form whatever constitutes the complete thing *simpliciter*, in the *ratio* of substance, for any form whatever gives *esse simpliciter* and substantial *esse*. Therefore, one form is not a disposition to another unless the substantial unity of the composite be destroyed. Substantial *esse* is not multiple. Therefore, substantial form cannot be multiple.[48]

IX. – Form is Educed from the Potency of Matter. One thing can be educed from another in three ways: (1) as from a place, in the manner in which money is drawn out of a purse; (2) as from an efficient cause, as an act is educed from a faculty; (3) as from a material cause in whose passive potency it is contained. Hence, to educe form from the potency of matter is to make it be in act in the matter in whose passive potency it was pre-contained, with that matter serving as material cause for the coming to be of the form.

1st Proof (Indirect Proof). To explain the production of a form, only four hypotheses can be conceived. The form either (a) was in act,

---

[47] *Summa theologiae* Ia, q. 76, a. 4: "[U]nius rei est unum esse substantiale. Sed forma substantialis dat esse substantiale. Ergo unius rei est una tantum forma substantialis."

[48] The question will again be treated in Psychology (IIa-IIae), where we shall speak of the form of corporeality in man. In the meantime, one can refer to Nys, *Cosmologie*, n. 135ff and 253ff.

but as if hidden, in matter, or (b) was extracted out of nothing, or (c) was educed from matter as from an efficient cause, or (d) was educed from matter as from a subject in whose passive potency it was contained. But it is all too absurd to imagine that forms are in act in matter as in a hiding-place; for if they were in act in matter, then the composite would already be constituted in act, and there could not be substantial generation. Second, form was not created, since a finite agent cannot draw anything out of nothing; and moreover, that which is created is not a form, but a composite. Third, it was not educed from matter as an efficient cause, for matter is merely passive. It remains, therefore, that it is educed from matter as from a subject in whose passive potency it was contained.

2nd Proof (Direct Proof). Three conditions are required for a form to be educed from the potency of matter: (a) that the subject be proportioned to the form and the form to the subject, otherwise the form would not be contained in the subject; (b) that the form depend on the subject in its *esse* and its becoming (*fieri*); (c) that the form do not come from the outside, but that it may arise from the matter itself, and that it come to be through the transmutation of matter. And these three conditions are verified in the case of non-subsistent forms. Therefore, non-subsistent forms are educed from the potency of matter. *The major premise* explains what it means to be educed from matter. *Proof of the Minor.* By non subsistent form we understand a form that has material *esse*. Now, that which has material *esse* is clearly proportioned to matter. Therefore, the form and the subject are proportioned to each other.

And so we have our first condition. But if the form has material being, it depends on the matter for its *esse*. But becoming is commensurate with being (*fieri esse commensuratur*). Therefore, the form depends on matter in its becoming. Thus, the other condition is verified. Now, we have already shown that form does not come to be out of nothing, that it is not infused by an extrinsic agent through creation. Therefore, it must arise through the transmutation whereby matter passes from potency to act, and thus we assign the third condition.

Objection: That which is not contained in act in matter cannot come to be in act in it. But form is not contained in act in matter. Therefore, it cannot come to be in act in matter, nor be educed from it. ***Reply***. *I distinguish the major.* That it cannot come to be in act in matter through the sole powers of matter, I concede; but that it cannot come to be in act in matter through the action of an agent, I deny. *I distinguish the conclusion.* That, therefore, it cannot be educed from matter through the sole powers of matter, I concede; but that it cannot be educed from matter through the action of an agent, I deny. The agent, by changing matter, certainly makes that which was only in potency be in act.

You will insist. But the action of the agent is an accidental form. Therefore, it cannot educe a substantial form. ***Reply***. *I distinguish the antecedent*: that an action is an accident that acts by virtue of the substantial form, I concede; but that it acts by virtue of itself, I deny. *And I deny the conclusion.* It is not at all contradictory that the accident, as an instrument of the substantial form, produces a substantial form, for then the efficacy of the action proceeds from the substantial form and is ascribed to it.

As is evident, we are concerned with material forms, for a subsistent form, since it has *esse* that is independent from matter, cannot be educed from the potency of matter. Thus, in the first production of things, form was not educed, because that would imply that form existed without matter; but matter and form were created simultaneously for the creation of the whole.

X. – **On the Drawing Out of Accidental Forms**. Accidental form can be either a property or a predicable accident. But properties are not educed from the potency of the subject. For the form that comes to be for the production of the whole is not educed from the potency of the whole, but is simultaneously with it. But the properties come to be for the production of the subject and emanate from the subject by virtue of that action whereby the subject is produced. Therefore, they are not educed from the potency of the subject; but, if the subject is created, then the properties are created simultaneously; if the subject comes to be by being educed, or through generation, then the properties are educed or generated simultaneously.

Now, all accidental forms that are not properties—that are necessarily connected with the subject—are educed from the potency of the subject.

Proof. For forms to be educed from the potency of a subject it is required—and suffices—that they presuppose a subject that has been already produced by an action, and that they depend on it in their becoming (*fieri*), in their *esse*, and in their being conserved (*conservari*). But the aforesaid forms, since they are not properties, do not emanate by virtue of the same action whereby the subject is produced; since they are *beings in act*, they depend on the subject in *esse*, and thus in becoming, because *esse* and becoming are correlative; and similarly they depend on the subject in being conserved, which is *esse* itself continued. Therefore, the said accidental forms are educed from the potency of the subject—a natural potency, of course, if they are natural forms, like knowledge, virtue, color, etc.; but an obediential potency if they are supernatural forms, like grace, charity, and all the infused virtues. For supernatural accidental forms are not created, because being created is proper to subsistent things; otherwise they exceed the potency of the natural subject. Therefore, they are educed from the obediential potency. An obediential potency is the very entity of the subject with respect to a superior agent that can induce an effect that exceeds the powers of nature.[49]

XI. – Form, Insofar as it is Of Itself, is Communicable. Form is the principle of the specific degree of being of a thing. But species, of itself, is predicable of many things and can be present in many things univocally. Therefore, form of itself is communicable. But communication can be impeded, not by a defect of the form, but by a defect of the receiving subject. For, in order for the multiplication of forms to occur, there must be many subjects; but it could happen that there is only one subject and then the species is preserved in only one individual. Another requirement is that there be a principle of individuation. But certain beings, namely angels, lack a principle of individuation, which is signate matter. Hence, the angelic form, even though it is communicable in itself as species, is not in fact communicated, because there are no numerically distinct subjects that can receive it. Moreover, the axiom is understood of the informing form, which is multiplied by reason of the material subjects.

XII – Form is Not Prior to Matter in Time, but is Prior in Dignity and Knowledge. It is not prior in time, because form cannot naturally exist without form; in the first origin of things, matter and form were created simultaneously, but in other productions the form is educed

[49] Cf. *Metaphysics: Ontology*, Treatise I, q. 2, a. 1.

from the potency of matter, as was just explained. But it is prior in perfection and dignity, for it is the form that gives *esse* and act; hence, matter is for the sake of the form, and not the form for the sake of the matter. Form is also prior in knowledge, for it is known and defined of itself, whereas matter is knowable and definable only through its relationship to form.

XIII. – The Last Form is a Property of the Appetite of Matter. Since the form is essentially incomplete, it is necessarily ordered to the matter, and it cannot naturally remain without matter. Hence, just as it seeks its own existence, so it seeks its union with matter.

This discussion gives occasion to the question of whether a non-subsistent form can, by a miracle, be separated from matter. There are two opinions. The Thomists answer in the affirmative, and many non-Thomists deny it. We chose the first opinion. Therefore, form needs matter because, since it does not subsist in itself, it requires matter as support. But God can supply the sustainment that prime matter provides, as He supplies the function of the substance in the Eucharist. Therefore, the form can be divinely conserved without matter, as an accident without a subject.

Objection: Form has one undivided *esse* together with matter. Therefore, it cannot exist without it. ***Reply***: Form is essentially incomplete and can only be individuated by being ordered to matter, and once individuated it can, by divine intervention, persist in its individual being, as quantity, once individuated, can be conserved without a subject in the Most Holy Sacrament of the Eucharist.

XIV. – On the Third Principle of Becoming, Namely, Privation, two things must be had in mind. First, privation does not coincide univocally with matter and form in the *ratio* of a principle, but only analogically. Of course, *ens* and *non-ens*, the positive and the negative, cannot coincide in the same univocal *ratio*. But privation is *non-ens* and a negative principle, whereas matter and form are real and positive principles. Therefore, privation does not coincide univocally with matter and form in the *ratio* of principle. Second, in matter there is not only one privation, but as many privations as forms are lacking. For privation is the absence of form in a capable subject. But there are as many absences of forms as there are absent forms, as is clear. Therefore, in any matter there are as many privations as there are forms that it lacks.

On natural composites

I. – The Nature of the Composite. A composite, which is said to be the terminus of generation, results from the union of matter and form. There are many things, therefore, that remain for us to discuss.

The opinion of Averroes and of many ancients concerning the nature of the composite was that the form is the whole essence of the natural being and that matter did not pertain to the intrinsic constitution of the composite any more than the subject of accidents pertains to their intrinsic power and *ratio*. The opinion of Plato and that of certain moderns approaches this view, insofar as they claim that the soul alone pertains to the essence of man, on which account man is defined as 'an intelligence using a body'.

II. – Conclusion: "Matter, and not only form, also pertains to the intrinsic *ratio* and constitution of the natural composite."

1st Argument. That which pertains to the intrinsic *ratio* of a physical being is expressed by its definition. But a definition in natural things does not include form alone, but both form and matter. *Therefore. Explanation of the Major.* This is the difference between a substance and an accident, that in the definition of accident is included not only that which it is, according to its intrinsic *ratio*, but also something extrinsic, namely, the subject; but, since substance is a complete, *sui juris*[50] *ens*, in its essential definition nothing extrinsic is to be included. *The minor premise is clear from the facts.* For man is not defined as 'that which consists of a soul', but, 'that which consists of a soul and a body'. Nor is man defined as 'rational', but as 'rational animal'.

2nd Argument. A composite is a generable and corruptible being: for when it arises, it does not come to be out of nothing; when it ceases to be, it does not fall into nothingness. But form without matter is neither generable nor corruptible. Therefore, form without matter is not

[50] *Sui juris*, in its own right. – ***The Translator.***

a composite. *Proof of the Minor.* Form is simple. But generable and corruptible being necessarily lacks simplicity; it requires at least two elements: one which is lost through corruption or that is acquired through generation, and another that generation presupposes and that corruption leaves behind. *Therefore.*

– ***Objection***: If common matter belongs to the nature of physical being, then for the same reason singular matter will belong to the essence of an individual. But the consequent is false. Therefore, the antecedent is false. *Proof of the minor*: That does not pertain to the intrinsic essence of a thing which can vary without the essence varying. But the matter of Peter continuously varies without his essence varying. Therefore, it does not pertain to his essence. – ***Reply***: *I concede the major, but I deny the minor*. I shall prove it by distinguishing the major. *I distinguish the major*: [if it means] that 'that which does not pertain to the essence *formally* varies, without the essence varying', I concede; but [if it means] that 'that which does not pertain to the essence varies only *materially*, without the essence varying', I deny. *I contradistinguish the minor*: that the matter of Peter varies *materially*, I concede; but that it varies *formally*, I deny. *And I deny the conclusion.*

Explanation of the Solution. The matter of Peter varies *materially* insofar as by virtue of daily work something from it is lost and something else is gained through nutrition; but *formally* it remains the same, because the supervening matter substitutes the preexisting matter within numerically the same composite, and it is designated[51] by numerically the same quantity and form, and thus the same numeric unity that the lost matter possessed is obtained. This is the opinion of the Thomists.

III. – Whether the Composite is Distinct from its Matter and its Form. It has been established that the composite results from the union of matter and form and that it is distinct from both the matter and the form taken separately; but the Scholastics additionally inquire whether the parts, taken collectively and joined together, are distinct from the whole itself which they compose. All concede that the composite is *virtually* distinct from the matter and the form taken together; but whether they are *really* distinct is disputed even among

[51] For more on the 'designation' on matter, see below, Treatise II, q. 4, "On the Principle of Individuation." – ***The Translator.***

Thomists. Scotus, Capreolus, Cajetan, and Ferrara are cited as being among those who answer in the affirmative. The Complutenses, John of St. Thomas, Goudin, Guérinois, Mailhat, and many others answer unanimously in the negative. We will adhere to the latter opinion.

The same thing is not really distinct from itself. But matter and form taken collectively are the very composite itself. Therefore, the composite is not distinct from matter and form taken collectively. *Explanation of the Minor.* The term 'composite' does not add any new concept to those of matter and form, for neither can there be found in the composite any substantial entity that is not included in the matter or the form. Therefore, matter and form taken collectively are the very composite itself.

– ***Objection***. Cause and effect are really distinct from each other. But matter and form are the cause of the composite. Therefore, they are really distinct from it. – ***Reply***. *I distinguish the minor.* That they are causes of the composite taken separately, I concede; but that they are causes of the composite taken collectively, I deny. For taken thus they are the composite itself. *I distinguish the conclusion.* That, therefore, they are distinct from the composite if they are taken separately, I concede; but that, [therefore, they are distinct from the composite] if they are taken collectively, I deny.

Further, as Goudin notes,[52] the disagreement among authors is more about words than about things. For those authors who seem to say that the whole is a third entity distinct from the matter and the form should be interpreted as thinking of matter and form as taken singly even in their union, and not as taken to be one thing through their mutual conjunction.

IV. – The Single Existence of the Composite. Just as the essence of the composite is not a certain third entity distinct from matter and form, so the existence of the composite is not distinct from the existence of matter and form; but one and the same existence belongs to the form, the matter, and the composite. This is the opinion of the Thomists, which is opposed to that of Scotus, Suárez, and others.

Proof. Existence is the terminus of becoming, or the terminus at which the thing is said in fact to be. But becoming and its terminus are proportional. Therefore, existence belongs to something in the same

---

[52] See *Phys.* Ia P., disp. 1, q. 4, a. 3.

way that becoming belongs to it. But becoming belongs to the composite as *that which* (*quod*), and to the parts as *that whereby* (*quo*). Therefore, the one existence that affects the composite renders the parts as existing, which exist, not of themselves, but only insofar as they are in the whole, by reason of which they coexist (*coexistunt*) and exist in (*inexistunt*) [the composite], as John of St. Thomas teaches.[53] *Proof of the minor*, namely, that becoming belongs to the composite as *that which* (*quod*). For becoming has the singular and the individual as its terminus. But form is not the individual itself; on the contrary, the composite is in itself individuated and subsistent. Therefore, the composite is what becomes primarily and *per se*, or what is properly generated, and not the form. Hence, the composite is properly *that which* (*id quod*) has being, whereas form is *that whereby* (*id quo*) something has *esse*.

Objection: The entity of the composite is composed. But a composite entity cannot have as its terminus a single and simple existence. Therefore, the existence of the composite is not single and simple. ***Reply***: *I deny the minor*. For just as all the parts that compose a whole have a single hypostasis or subsistence as their terminus, and all the parts of a line have a single point as their terminus, so all the parts of a composite essence can be actualized by a single existence; existence, of course, is the last terminus, and the last actuality of every thing, and, although indivisible, it is nonetheless multiple virtually and extensively.

V. – The Immediate Union of Matter and Form. The existence and essence of the composite are immediately united without an intermediary mode. Is the same to be said of the union between matter and form when they constitute the composite? The opinion that claims that the union is something superadded to matter and form, namely, a certain substantial *mode* whereby form unites to matter, was formerly quite common. This was the opinion of Scotus, Durandus, Suárez, and many Thomists, even if not all of them explained the matter in the same way. The more common opinion, however, holds that matter and form are immediately united. It must be noted that we speak here of the formal union, for, if our discussion were only concerning the effective union, it is certain that matter and form are united through the action of the agent, as we previously mentioned. Omitting the other arguments that defend the second view, we provide this one:

---

[53] See *Phil. nat.* Ia P., q. 7, a. 5.

Argument. Matter is essentially potency, and form is essentially act. But it is altogether necessary that potency and act are immediately united. *Therefore. Proof of the Minor.* If between potency and act there were an intermediary mode, then this mode must again be either potency or act. Therefore, if act and potency are not immediately united, then the aforesaid mode must again be united through something else, and thus there will be an unending process.

Confirmation through the Authority of St. Thomas: "Form through itself makes matter be in act, since its essence is act. *Nor does it give being through some medium*, but according to itself it is united to matter as its act. Nor is there something that unites, except the agent that makes matter be in act."[54]

1st Objection: That which ceases to be while something else remains is really distinct from that which remains. But, the union ceases to be while matter and form remain. Therefore, the union is something distinct from them. ***Reply***. *I distinguish the minor*: that the union ceases to be while matter and form remain as mutually communicated, I deny; but that the union ceases to be while the matter and the form remain as separate, I concede. *I distinguish the conclusion.* That the union is something distinct from them taken separately, I concede; but that it is something distinct from them as mutually communicated, I deny. The union is neither the matter alone nor the form alone, but matter and form as mutually communicated. But when they are separated, a third reality does not cease to be, but rather the form stops being sustained by matter and the matter stops being actualized by such a form.

2nd Objection: Union is a certain causality of matter and form. But causality in creatures is distinct from the cause. Therefore, union is distinct from matter and form. ***Reply***. *I distinguish the major*: That union is an intrinsic causality, I concede; but that it is an extrinsic causality, I deny. *I contradistinguish the minor*: That causality is distinct from the cause among extrinsic causes, I concede; but that they are distinct among intrinsic causes, as are matter and form, I deny. For their causality is their entity itself as communicated.

---

[54] *Summa theologiae* Ia, q. 76, a. 7: "Forma autem per seipsam facit rem esse in actu, cum per essentiam suam sit actus; *nec dat esse per aliquod medium*... [sed] secundum seipsam unitur materiae ut actus eius. Nec est aliquid aliud uniens nisi agens, quod facit materiam esse in actu."

VI. – The Composite With Respect to the Accidents. The question is raised whether the composite is the subject of the inhesion of the accidents. Three hypotheses are conceivable: The subject of accidents is either: a form, prime matter, or the composite. A subsistent form is a subject of spiritual accidents, as are intellect, will, thinking, and volition, which cannot be received in a material thing. With respect to material accidents, however, all concede that the form is not the subject, because it is simple, whereas accidents participate in a certain kind of extension; and moreover, the form has the *ratio* of something that actualizes, not that of something that receives. There remain, therefore, two opinions: The first claims that the subject of accidents is prime matter, either according to itself or supposing the presence of a substantial form; this is the opinion of Suárez and the Conimbricenses. The other opinion asserts that only the composite can be the subject of the inhesion of accidents. This is the opinion of Capreolus, Cajetan, Bañez, John of St. Thomas, the Complutenses and unanimously of the Thomists.

Proof of the Latter View. The proper subject of accidents requires two conditions: (1) it must exist in itself, and (2) it must give *esse* to accidental forms. But prime matter possesses neither of these qualities. Therefore, it is not the proper subject of accidents. *Explanation of the Major.* The first subject of accidents must exist in itself for, of course, the act of subsisting in itself is prior to the act of underlying something else. Second, it must give *esse* to accidents. For this is the difference between substantial and accidental forms, that the substantial form gives its first *esse* to the subject and makes it be in act *simpliciter*, whereas accidental form requires a subject in act and depends on the very *esse* of the subject as on the first *esse* that is presupposed. Therefore, the subject gives *esse* to accidental forms. *The minor premise is clear.* To exist *per se* and to confer *esse* to others belongs to the composite, and in no way to matter, which according to itself includes no act of essence or of existence (*nullum actum essentiae vel existentiae*).

1st Objection. Although *de jure* matter does not subsist, nonetheless *de facto* it is not without form; for this reason it subsists *de facto*, and can underlie accidents.[55] – ***Reply***: Even if in fact matter is not without form, it does not subsist by virtue of itself, but only insofar as it

[55] *De jure*, in principle (lit. 'by right'); *De facto*, in fact. – ***The Translator.***

is in the composite and by reason of the composite. Therefore, it is only a subject by reason of the composite.

*2nd **Objection***: Quantity is a subject of accidents. But quantity is an accident of prime matter. Therefore, accidents are sustained by prime matter. ***Reply***: Quantity can be said to be the proximate subject of accidents, because it is a disposition for accidents; thus, whiteness, heat, etc., belong to substance only by means of quantity. But yet quantity is not the first subject, as if it supplied the *ratio* and radical power of sustaining something; rather, it itself is sustained by something else. With this in mind concerning the major premise, we can distinguish the minor premise. *I distinguish the minor.* That quantity is an accident of prime matter, that is, something that results in the composite by reason of prime matter, I concede; but that it is an accident that is directly sustained by prime matter, I deny. *And I deny the conclusion.*

Matter is the principle and the root of quantity, and for this reason quantity is considered an accident of matter; but yet matter does not have quantity in act, nor can it sustain quantity, since it does not subsist. Just as quality, even though it is said to be an accident of form, does not thereby have form as its subject, so in the same way, even though quantity arises in the subject by reason of matter, it is not received in matter as a subject.

VII. – When the Composite Ceases to Be, All Accidents Are Destroyed. This is a corollary of the preceding. Since an accident cannot exist independently of a subject, except through a miracle, nor can it migrate from subject to subject; when the subject ceases to be, the accidents cannot remain formally the same. But the subject of all material accidents is the composite whole, not bare matter. Therefore, when the composite ceases to be, the accidents do not remain numerically the same.

*1st **Objection***. From the facts of chemistry it is clear that the same weight remains [after a chemical reaction]. Therefore, at least the quantity does not change after the composite ceases to be. ***Reply***: What remains is the same weight, the same quantity *materially*, but not *formally*. For, since weight and quantity arise in the composite by reason of the matter and the matter remains, they remain materially the same; but since they are sustained in a new composite, they are formally distinct. Chemistry, moreover, can only give evidence of material identity or diversity; it pertains to philosophy to investigate formal identity and diversity.

2nd ***Objection***: If the accidents of a corpse are not the same in number as those of the living animal, then the senses are deceived insofar as they report that the figure, color, etc., are the same. ***Reply***: The reply given previously stands: the senses attest that it is the same figure, the same color, etc., *materially*; reason, however, argues that figure, color, etc., are *formally* different, since they have a new subject; and the senses do not in any way contradict this assertion.

3rd ***Objection***: A cause is to be assigned that reproduces the same accidents in the composite. But no such cause can be assigned. Therefore, there remain the same accidents as before. ***Reply***: *I deny the minor*. The same cause that destroys the first composite and produces the new composite also produces the new accidents. The dispositions that were in the previous composite possess an order to accidents that are similar to the previous accidents, and in the matter there remain certain impressions in virtue of which these accidents, rather than those, must arise in the composite, and the molecules retain the same structure, shape, etc.[56]

VIII. – THE COMPOSITE IS THAT FROM WHICH ACCIDENTS EMANATE AND ARE EDUCED. We dealt with this question where we discussed the drawing out of accidental forms. Proper accidents emanate from the composite as a natural result. For those things that have a necessary connection with a substance result from the very action whereby the substance is produced. But proper accidents have a necessary connection with the composite. Therefore, they result from the same action whereby the composite is produced. – Predicable accidents, however, are educed from the composite, as was explained in the same discussion.[57]

IX. – FROM THE FOREGOING THE TRUE NOTION AND DEFINITION OF 'BODY' IS GATHERED. A physical body, in the scholastic view, is the natural composite itself, as we have described it. Hence, it is easy to refute the diverse opinions of philosophers concerning the essence of a body. Gassendi claimed the essence of a

---

[56] On this point one may refer to NYS, *Cosmologie*, nn. 241ff.

[57] Above, a. 3, n. 10.

body was impenetrability, or solidity; Descartes claimed it was three-dimensionality; Locke, a collection of accidents; more recent philosophers, a heap of many substances.

But impenetrability, insofar as it is an effect of quantity, far from constituting an essence, presupposes both essence and quantity. Descartes confused the notion of 'natural body' with that of 'mathematical body'. Further, quantity and three-dimensionality are distinct from a physical body, as property is distinct from essence, as we shall show below.[58] Moreover, a physical body possesses many properties that cannot in any way be deduced from its extension, as are its shape, its powers, etc. Therefore, extension and three-dimensionality cannot express the genuine and adequate essence of a body. Locke perverted the true notion of substance. Only something that is *per accidens* one results from a heap of substances. Therefore, if he were correct, then substance would not be *per se* one. Further, each accident needs a foundation that underlies it; therefore, even heaped together, they require sustenance, and on that account they altogether necessarily presuppose a substance. The opinions of more recent authors are rejected from the refutation of atomism and dynamism. Of course, a corporeal substance must be *per se* one. But from many substances that are united, only something that is *per accidens* one comes to be. Therefore, a body is not an aggregate of many substances.

We concede, however, that the essence of body requires three dimensions. For, since matter is the root of quantity, there cannot naturally be a body that is not extended. Quantity is the first accident that inheres in the essence and by means of which the other accidents are received; hence, matter has its quantity first before receiving its form. Extension, therefore, and three-dimensionality are naturally present in a natural body. All of these things will become clearer when we discuss quantity and the principle of individuation.

There are, therefore, two definitions of body: namely, a descriptive definition and an essential definition. A body is defined descriptively as: "a substance that requires three dimensions in space" (*Substantia trinam exigens dimensionem in spatio*). This definition is better known to us (*quoad nos*), for we observe extension and properties of a body before its intimate quiddity. And the essential definition is: "a substance composed of matter and form" (*Substantia composita ex materia et*

---

[58] Q. 3, a. 1.

forma). This latter definition is established from the foregoing, especially from no. II of this article.

Indeed, body can be absolutely defined as: "A substance that is essentially composed [out of matter and form]" (*substantia essentialiter composita*). Of course, the only substance that can be essentially composed is that which consists of matter and form. For the composition of essence and *esse* is a metaphysical composition, not an essential composition, for *esse* is really distinct from essence; hence, the terms *essential* composition, *physical* composition, and composition *out of matter and form* are used indiscriminately. But so that equivocation be removed and the definition be understood upon first consideration, the words "out of matter and form" are skillfully added, even if they are somewhat redundant. In practice, therefore, we shall indifferently name 'body' any substance that is essentially composed or any substance that consists of matter and form.

Explanation of many things that are necessary for the complete understanding hylemorphism

I. – *Ratio* of the Article. Having vindicated the scholastic view on the essential constitution of bodies, we add many things to complete this doctrine, especially concerning substantial generation, which is the foundation of the whole system of hylemorphism.

II. – Generation, as we have explained many times, is the production of a substance from a preexisting subject. It differs from *creation*, which is the making of a substance, but not out of anything presupposed, and from *alteration*, which is the production out of something presupposed, but not the making of a substance. Now, there are four species of generation: generation *simpliciter* (*generatio simpliciter dicta*), mixing (*mixtio*), generation of elements out of a mixture (*generatio elementorum ex mixto*), and the generation of living things (*generatio viventium*).

Generation simpliciter occurs when a new substance is produced in such a way that the previous composite is totally destroyed and there only remains prime matter. There are two kinds of *mixing*: one type of mixing is called *'mixing according to sense'* (*mixtio ad sensum*), and it is an aggregation, or blending, of many bodies that retain their own nature. This is the kind of mixing that chemists call 'a mixture', and it occurs when the smallest particles are juxtaposed. Another kind of mixing, which corresponds to *chemical compounds*, can be defined, according to St. Thomas's words, as: "The union of mixable things that have been altered and corrupted according to forms."[59] The Angelic Doctor observes three things concerning mixing: First, elements transform each other, act upon each other, and react by mutual and contrary action and passion by reason of contrary qualities, and out of contrary qualities that are received in varying degrees is constituted an quality that is intermediate between the nature of each extreme. Second, the mixture differs in species from each mixable thing (for example,

[59] Anonymous, *In I De generatione et corruptione*, lect. 25: "Miscibilium alteratorum et corruptorum secundum formas unio." This is an anonymous continuation of Aquinas' commentary, once ascribed to St. Thomas himself. – ***The Translator***.

water differs in species both from oxygen and from hydrogen). Third, there must be a proportion between each mixable thing because, if they are too disproportionate, the species of one is broken up by the excess of the other.

— *The generation of elements out a mixture* is opposed to a mixture as to a contrary. For a mixture occurs through the synthesis of elements, as when water is produces through the synthesis of oxygen and hydrogen; but the generation of elements occurs through analysis, or through the resolution of a mixture into its elements, as for example when, out of the resolution of water, oxygen and hydrogen come to be. — The fourth kind of generation is the *generation of living things*, which is defined as: "The origination of a living thing from another living thing, from a conjoined principle, in a similitude of nature" (*Origo viventis a vivente, principio conjuncto, in similitudinem naturae.*) Certainly it is necessary that the generated thing be something living, and that it arise out of a living thing that generates it, through something that was in the generating thing and remains in the generated thing; and finally, it is necessary also that it coincide in the specific nature with the living thing in virtue of its own production. We shall discuss this again in Psychology. But here arises an often-repeated question on mixed composites.

III. – On the Permanence of Elements in the Mixture. So, what is the status of simple elements within a chemical compound, or in a mixed body? Element is defined according to Aristotle as: "That into which the bodies in which it is present (in potency or in act) can be divided and which is itself indivisible into diverse things according to species" (*Illud corporum in quod caetera corpora dividuntur, in quibus inest* (potentia vel actu), *ipsum vero est indivisibile in diversa secundum speciem*). Four things, therefore, belong to the *ratio* of element: (1) it is that out of which a body is composed, and so it pertains to the material cause; (2) it is the first thing out of which it is composed; (3) it is intrinsic to the thing, and hence a thing subject to privation or to contrariety cannot be an element; (4) it cannot be divided into other species.

This question on the permanence of elements in the mixed body is often repeated no less among Scholastics than among many recent authors. Three possible solutions are conceivable. Either elements remain *formally in act*, or *merely in potency*, or *virtually*. They are said to remain in act if they retain their own *esse* and their own forms in the composite. They would remain merely in potency if they were reduced to prime matter such that they would only retain a passive inclination to the form and to the *esse* that they previously possessed. And they would remain virtually if in the subject is preserved a certain active inclination for this form to be actuated rather than another, and if properties are reproduced in the new composite which are similar to the properties that existed previously.

The first solution is commonly (*communiter*) ascribed to the more recent scientists and chemists[60]; the second is not favored by any distinguished philosopher; the third is more common (*communior*) among Scholastics.[61]

IV. – CONCLUSION: "In chemical compounds, or mixed bodies, elements remain neither formally in act nor merely in potency, but virtually."

Proof of the 1st Part. 1st Argument. If elements remained in act according to their substantial form, there would be many substantial forms in one composite: that is, as many as there are elements. But we already have shown that it is impossible for there to be many forms in the same composite. Therefore, it is impossible for elements to remain in act.

---

[60] But perhaps scientists, who engage in a different order of discourse, understand the issue in a different sense. For they, who pay attention only to proximate causes, consider especially the very *fact* of permanence. They do not care whether this permanence is *formal* or *virtual*; it is an extraneous issue that they relegate to the philosophers. Further, scientists assert this permanence *for the instant at which elements are composed through synthesis,* or *at which they are dissolved through analysis*; whether they remain *formally* in the stable state of compound is again an issue extraneous to chemical science and is to be considered on the basis of other principles.

On this account, the question of the permanence of elements in the mixture, just like the whole question of matter and form in general, which we already discussed, is not a dispute between Scholastics and scientists, but rather a metaphysical problem to be dealt with by philosophers among themselves, although the natural sciences are especially relevant.

[61] This question is treated anew by D. NYS, *Cosmologie*, n. 254ff. Cf. GREDT, t. 1, pars 2.

2nd Argument. Two facts from philosophy and chemistry can be proven: (1) The composite that arises out of the combination is altogether homogeneous; (2) it possesses properties that are specifically distinct from those of the components. But, if elements remained formally in the composite, neither fact would be explained. *Therefore.* – The first fact would not be the case. – For elements either penetrate each other or remain juxtaposed. Further, the mutual penetration of bodies is a miracle; and if they remain juxtaposed, then no one molecule of water is water, but one would be oxygen, the other hydrogen. And so bodies would not be homogeneous. Nor would the second fact be the case. For elements to remain in act means for it to persevere according to the same *esse*. But properties follow *esse*. Therefore, if elements remain in act, then their attributes persevere altogether, and on that account it is impossible to explain the specific diversity of properties.

Nor can it be replied that this can be sufficiently explained if we say that the opposite properties of the elements are neutralized. For, if they are neutralized, there would be no effect; nor can one reasonably conceive how mere neutralization can cause specifically diverse properties.

Whatever the adversaries offer as a middle term for their argument, they are always defeated by the same argument: specifically diverse properties argue for specifically diverse being; the composite that is *per se* one cannot receive many substantial forms.

Father Munnynck, O.P., explains the issue both scientifically and philosophically.[62] Among other things, he teaches that atoms do not remain formally in the molecule; that is, that they do not retain their *essential* distinction; but he says that, nonetheless, they retain a certain *accidental* (e.g., quantitative) distinction.

This opinion, which is so excellently set forth by the author, is useful for students to know; they should read what the author himself discusses.

Here we only observe that distinction which says that the *esse* of the composite remains only materially and not *formally.* For indeed, when the subject ceases to be, or does not retain its essential unity, accidents no longer remain numerically the same. But an atom, as it is supposed, does not retain its specific *esse*, or its essential unity.

[62] See his *Notes sure l'hylémorphisme.*

Therefore, the accidents of atoms, or their accidental distinctions, are not the same formally, but only materially.[63]

Proof of the 2nd part. Through the analysis (or separation) of a compound the same elements always return. But, if they remained within the compound only in passive potency, since that potency is indifferently related to all forms, there would be no reason why the forms of elements constantly and invariably reappeared. Therefore, elements persevere in the composite according to a certain mode that is intermediary between act and purely passive potency, that is, virtually. Hence, the third part of the conclusion is proven from the first and second parts.

V. – Difficulties Resolved.

1st Objection. Chemists can associate or dissociate elements as they desire. Therefore, elements exist in act in the composite. ***Reply***: That they exist in act *in that instant* at which they are associated through synthesis or dissociated through analysis, I concede; but that they exist in act in the very *state* of being a compound, I deny. – Only two things can be concluded from the chemical facts: that elements are in act in the instant of composition or resolution, and that they persist virtually in the state of mixed body. But the question of whether they also remain formally or not is to be resolved on the basis of other principles. ***You will insist.*** They remain in act. Often heat, light, electricity, or mere friction is enough to produce a chemical compound. But that can only be explained if elements persist in act. *Therefore.* ***Reply.*** *I deny the minor.* Physical agents, of course, such as light, electricity, etc., move the powers of elements that are *radically* in the previous composite and which persist in the same; powers thus excited pass to be *in act* and, thus, chemical combining is explained, without us being compelled to admit the actual permanence of elements in the mixed body.

2nd Objection: Where the properties of the thing remain, there the form remains. But the properties of the elements, such as their quantity, persist in the mixed body. Therefore, its form also remains. ***Reply***: *I distinguish the major*: that where the properties that belong to the thing by reason of its form remain, there the form remains, I concede; but that where the properties that belong to the thing by reason of the matter, there the form remains, I deny. *I contradistinguish the minor*: [That the properties that belong to the elements by reason of the matter remain in

[63] Cf. a. 3, nn. 6-7.

the composite, such as their quantity, I concede; but that the properties that belong to the elements by reason of their form remain in the composite, I deny.] *And I deny the conclusion.* – We have already explained how quantity arises from matter and how it remains the same materially after the coming to be of the composite.[64] ***You will insist***. The powers that belong to the elements by reason of the form remain. For affinity remains, and affinity belongs to the elements by reason of the form. *Therefore.* ***Reply***. It is gratuitous to assert that affinity remains in act; for, in order to explain chemical experiments, it is sufficient that it persists virtually and in likeness. By the very fact that they persevere in a hidden and attenuated matter, they can be excited through the affinities that correspond to them and pass to act. Now, it has been established that the new composite comes to be only when the previous composite is destroyed; from this we argue that affinities are in act only when the mixed body is resolved and destroyed.

3rd Objection. From spectral analysis it has been established that the colors of the simple elements appear in the composite. But this fact shows that the powers of elements remain in act in the composite. *Therefore.* ***Reply***. *I distinguish the major.* That the colors of the elements appear in the composite while the composite remains at rest in the compound state, I deny; that these colors appear while the mixed body begins to be resolved through the action of light or heat, I concede. *I contradistinguish the minor*: that this fact shows that there are powers in act in the mixed body, if this happens in the compound state itself, I concede; but that this shows there are powers in act in the mixed body if this happens only when the mixed body begins to be resolved and destroyed, I deny. *And I deny the conclusion.* That only implies that the powers persist in similar entities.

For the colors of elements to appear, a vehement action is required whereby the mixed body begins to whiten and burns, whether through the action of limelight or through an electric influx. The mixed body, therefore, is no longer at rest in the compound state, but is somewhat being resolved into its elements. Further, we do not object to the idea that the elements are in act at the instant of resolution or of forming a compound. The issue is merely whether they remain formally in the compound state at rest, and the aforesaid phenomena do not prove this. Further, arguments deduced from spectral analysis are not very cogent,

---

[64] See previous article, n. 7.

for the aforesaid colors do not appear complete and distinct, but modified, and exhibit new particularities.

Thus, the natural sciences do not cite any certain facts against our conclusion. Indeed, many experts in chemistry fall back upon the scholastic doctrine through another process of reasoning. Berthelot says that, "It has been demonstrated, then, that the composite was truly in potency," (and we must understand this as meaning virtual potency), "with all its qualities in the component bodies evidenced through [spectral] analysis."[65]

Rubbini says, "To prove that elements retain their nature in the composite without undergoing change in their substance, we would need to demonstrate the actual existence of those elements in the composite, such that it remains what it is, that is to say, until the moment of its decomposition; and this fact has not been demonstrated, whether by direct observation, or by any rigorous form of reasoning founded on the observation of facts, independently of all hypotheses."[66]

Most recently, P. Duhem has said:

> "For Aristotle, all philosophical research has as its foundation a very meticulous and precise logical analysis of concepts that arise in our intellect through perception. Is the issue, for example, the philosophical problem of mixed bodies? Then one must, before all else, bring about that which an exact analysis distinguishes in this notion: elements that cease to exist at the moment when the mixed body is created; a homogeneous mixture whose smallest parts enclose within themselves the elements in potency and can regenerate them through their own corruption. The imagination of the atomists substitute hypotheses about the persistence and juxtaposition of atoms for those characters that are necessary and sufficient for constituting the notion of the mixed body. Those hypotheses, whose objects are not at all knowable by our legitimate means of knowing, *should be pitilessly relegated to the realm of chimeras*. Current physics, too, places as the foundation of all theory an exact logical analysis of concepts with which experience

---

[65] BERTHELOT, *Synthèse chimique*, p. 7: "Il est donc démontré que le composé se trouvait réellement en *puissance* avec toutes ses qualités dans les corps composants mis en évidence par l'analyse."

[66] See Farges, MATIÈRE ET FORME, IIe partie, VI.

furnishes us; through this analysis, it strives not only to mark with precision the essential elements that compose each of those notions, but also carefully to eliminate all the parasitic elements that mechanist hypotheses have gradually introduced into them."[67]

VI – On the Causes of Generation. Now that we have explained what generation is and what is the role of elements in the generation of mixed bodies, we must turn to the causes that prepare for or carry out the generation. Now, the preparatory causes are called dispositions. Further, it is quite obvious that dispositions are required for generation to occur, for generation is a substantial change in virtue of which a determinate form comes into matter. Therefore, certain prerequisites that allow the matter to change must concur with the causes that introduce the form into the matter. Hence, through experience we see in food that certain dispositions antecedently concur for the introduction of the form of flesh, and in the log we see that there are dispositions that are prior to the introduction of the form of fire.

VII – Alteration. The principal disposition that prepares for generation is *alteration.* In general, alteration means any accidental change, but here we take it in the special sense as a change that has a quality as a terminus, for quality, more than any other accident, makes something be otherwise than before. It is defined as, "A motion from a

[67] *Le Mixte*, p. 202-203: "Pour Aristote, toute recherche philosophique a pour fondement une analyse logique très minutieuse, très précise, des concepts que le perception a fait germer dans notre intelligence. S'agit-il, par example, de philosopher sur le mixte? Il faudra, avant tout, faire ressortir ce qu'une exacte analyse distingue en cette notion: des éléments qui cessent d'exister au moment où le mixte est engendré; un mixte homogène dont la plus petite partie renferme en puissance les éléments et peut les régénérer par sa propre corruption. A ces caractères nécessaires et suffisants pour constituer la notion de mixte, l'imagination des atomistes substitue des hypothèses sur la persistance des atomes et sur leur juxtaposition: ces hypothèses, dont les objets ne sont point saisissables à nos légitimes moyens de connaître, *il les faut reléguer impitoablement dans la région des chimières*. La Physique actualle, elle aussi, met à la base de toute théorie une analyse logique exacte des notions que l'expérience nous fournit; par cette analyse, elle s'efforce non seulement de marquer avec précision les éléments essentiels qui composent chacune de ces notions, mais aussi d'éliminer soigneusement tous les éléments parasites que les hypothèses mécanistes y ont peu à peu introduits."

quality toward another quality that is successively acquirable" (*Motus a qualitate ad qualitatem contrariam successive acquisibilem*). It is not a motion toward all acquirable qualities, but only into qualities of the third species, that is, sensible and tangible qualities, such as heat and cold, humidity and dryness. It occurs through a continuous motion and through successive changes, and in this respect it is different from generation, which occurs in an instant. The Scholastics distinguish three species of alteration: namely, alteration strictly speaking, intension, and remission. *Alteration* strictly speaking is a motion from one quality toward another that is properly and entitatively contrary to it, e.g., from cold to heat. *Intension* is the motion from an imperfect quality to the same quality in a more perfect [or more intense] state. And, finally, *remission* is the motion from an intense state of a quality to a less intense [or more remitted] state of the same quality. Further, 'intension' means the arrival of some superadded perfection in virtue of which the quality is made more actual and more efficacious; but intension cannot occur through the addition of a quality to a preexistent quality, as the Thomists teach unanimously. For if an addition of a quality to a quality occurred, then there would simultaneously be in the same subject two accidents that are only numerically distinct. But it is impossible for two accidents that are only numerically distinct to be in the same subject, as will be proven elsewhere, when we speak of the principle of individuation. Therefore, intension does not occur through an addition of a quality to a quality, but consists in a new mode of perfection by reason of which the quality becomes more rooted in the subject and actuates it more and perfects it more.[68]

Rarefaction and *condensation* are reducible to alteration, which are also dispositions for generation. For alteration is a motion that has a sensible quality as its terminus. But rarefaction and condensation have, as their terminus, sensible qualities, namely rarity and density, by virtue of which the same particle of matter can occupy various amounts of space. Therefore, rarefaction and condensation are reducible to alteration. The same is to be said of the many accidental changes that modern physics studies.

VIII. – Motion in Place, or 'Locomotion'. Among the prerequisites for generation we must include locomotion. For the agent

---

[68] We shall discuss these in *Ontology*, within the treatise on quality.

and the patient must exist together for change to occur, and those things are together which exist in the same place. But place acquires a body only through locomotion. Therefore, any change, whether generation or alteration, requires locomotion.

Confirmation. All change in quantity implies change in place. But all alteration requires change in quantity. Therefore, all alteration requires change in place. *The Major Premise is Established.* Quantity implies a relationship to a place; it has distinct parts in different places. Therefore, once a change in quantity occurs, there follows a change in place. *Explanation of the Minor.* All material accidents, quality not excepted, are immediately subjected to quantity, and by means of quantity, to a substance. On this account, once a change in quantity occurs, qualities also change, and vice versa; once quality changes through alteration, a change in quantity occurs.

IX. – WHETHER GENERATION IS ESSENTIALLY DISTINCT FROM ALTERATION. Many philosophers do not believe that generation is its own kind of action (*actio sui generis*), but that, once alteration occurs, generation follows as a result. Others, however, like John of St. Thomas,[69] the Conimbricenses, and Mielle, hold that generation is its own kind of action (*sui generis*) really distinct from alteration.

X. – CONCLUSION: "Generation is not the result of alteration; rather, a generated substance is produced through a true efficient cause distinct from alteration, the efficient cause whereby quality itself is produced."

Proof. Mutations that have diverse subjects differ essentially from each other. But alteration and generation require diverse subjects. Therefore, they differ essentially. *Proof of the Minor.* Alteration, like the other accidents, does not have as its subject bare matter, but rather the whole composite; but generation, on the other hand, has as its subject prime matter, which is defined as, "the subject of substantial change" (*subjectum mutationis substantialis*). Therefore, generation and alteration differ in subject.

Nor can a generated thing be said to be the result of alteration. For that which is a result of something else exists simultaneously with it and comes to be through the same action, and when that ceases to be, the

[69] JOHN OF ST. THOMAS, *De generatione subst.*, a. 3.

result also ceases to be: thus the properties that result from the essence exist simultaneously with the essence and cease to be when the essence ceases to be. But a generated thing is incompossible[70] with alteration at the same instant, for alteration is in the composite that is to be corrupted, which does not exist at the instant of generation. Therefore, the generated thing is not the result of alteration. For the generation of a new substance, therefore, a certain *sui generis* action is required that occurs in an instant, while alteration is a successive action.

XI. – The Efficient Cause of Generation. Those who hold that substance is immediately operative, like Scotus, Durandus, and Ockham, assert that substance is the efficient cause of generation; hence, they regard the following dictum as solemnly true: "Substance immediately generates substance" (*Substantia immediate generat substantiam*). But we already refuted this opinion.[71] Now, in the opinion of St. Thomas, which holds that substance acts though accidents, the efficient cause of generation is an accident, which acts in virtue of the substantial form, or, in other words, it is the very power of the generating thing as an instrument of the form. Indeed, the power of that which generates produces a twofold effect: a proper effect, namely, a preparatory disposition, which it produces as principal cause; and another effect which it elicits instrumentally under the influence of the generating cause and in virtue of which it attains the coming to be of the generated substance itself.

XII. – How the Last Disposition Relates to Generation. Among dispositions, some are *prior*, others *concomitant*. *Prior* dispositions precede in time the process of generation, have as their subject the composite that is to be corrupted, work for the composite's destruction, and perish simultaneously with the composite at the last instant of its existence. *Concomitant* dispositions are found in the generated thing at the very instant of its generation and in some way are dispositions for it. They are also called "last dispositions."

Objection. Here arises a great difficulty. If accidents had prime matter as their subject, then it would not be a problem to think that the dispositions of the subject that is to be corrupted will remain in the

---

[70] *Incompossible*, i.e., it cannot possibly exist simultaneously with something else. – ***The Translator.***

[71] Concerning dynamism, see arg. 4 in the preceding question.

subject that is to be generated; but we argue that the subject of accidents is the whole composite. Therefore, either the last disposition has as its subject the previous composite, and thus it perishes with the old composite and is of no use to the new composite; or it has as its subject the new composite, and then it is not a disposition for that new composite, since it is posterior to it. ***Reply***. We reply to this difficulty with the Thomists: The last disposition proceeds from the generated form as from an efficient cause and is received in the new composite, but it is a disposition for that form; hence, the form is posterior in the genus of efficient cause, but it is prior in the genus of dispositive and material cause.

Proof. Accidents proceed, as from an efficient cause, from the form of which they are passions and on which they depend in *esse*. But "the last disposition, e.g., heat in the eighth degree (*calor ut octo*) is an accident connected with the form of fire from that which is dependent in *esse* as a proper passion contained in its root, as we clearly see that generated fire has heat in the eighth degree as a property."[72] Therefore, the last disposition proceeds effectively from the generated form, and consequently it is received in the new composite.

And nonetheless it has the *ratio* of a disposition. All those things without which form cannot connaturally be received in matter have the *ratio* of disposition. But form cannot be connaturally received in matter without quantity and the other accidents. Therefore, the aforesaid accidents possess the ratio of accidents.

Further, that which is the terminus of preceding dispositions is itself a disposition to form. But the last disposition is a consummative terminus of the preceding dispositions. Therefore, it is a disposition to a new form.

Hence, the last disposition is the terminus of alteration and generation in diverse respects. It is the terminus of alteration insofar as it the consummation and terminus of the preceding alteration, but it is the

---

[72] JOHN OF ST. THOMAS, *De generatione*, q. 1, a. 7: "[U]ltima dispositio, v.g., calor ut octo est accidens connexum cum forma ignis ab illa que dependens in esse tanquam propria passio in ejus radice contenta, sicut clare videmus quod ignis genitus habet calorem ut octo tanquam proprietatem." *Calor ut octo*: this seems to be a reference to an early modern sort of temperature scale, according to which things ignite when they reach the eighth degree. –***The Translator.***

terminus of generation insofar as it is properly a passion connected with the substantial form, and thus it obtains the substantial form.[73]

Nor is it surprising that two causes precede each other in diverse genera of causes. Thus, in order to use a trite example, the wind, when it enters, actively concurs with the opening of the windows and is prior to it in the genus of efficient cause, and at the same time the opening of the windows concurs with the entering of the wind and is prior to it in the genus of quasi-material and dispositive cause.

XIII. – The Last Disposition does Not Fall within the Genus of Efficient cause. It has been shown that the last disposition concurs in the genus of material cause. But does it have also the *ratio* of efficient cause? The Thomists deny this. For the last disposition is received in a generated composite. But an accident cannot act effectively upon the subject in which it is received, because it requires that the subject already exist. Therefore, the last disposition cannot effectively cause generation.

XIV. – On the terminus of generation. There are four things that are produced in substantial generation: (1) substantial form, which comes to be in matter in virtue of generation, such as a corruptible form, or which at least is communicated to matter, such as the human soul; (2) the physical composite, or the essence of the whole that coalesces out of matter and form; (3) the metaphysical composite of essence, subsistence, and existence, which exists *per se* and incommunicably, and is called the supposit; (4) proper accidents also come to be, or at least result, in virtue of generation. And so it is asked whether all these things are the terminus of generation.

Now, we distinguish two senses of 'terminus': the adequated terminus, or the *terminus qui* ('terminus which'), and the formal or specifying terminus, or the *terminus quo* ('terminus whereby'). The *terminus qui* is the ultimate terminus of generation, that to which generation ultimately tends and which includes all the elements that exist in the final reality that is produced: the constituent nature, the substance whereby it subsists, and the existence whereby it exists. The formal terminus, or the *terminus quo*, is the immediate terminus of generation

[73] Cf. John of St. Thomas, *loc. cit.*

whereby the generated substance is constituted in its complete species, the terminus by means of which it attains the *terminus qui*.

Many opinions are offered on this matter. For some claim that the terminus of the whole is either the composite or the substantial form; others that the *terminus quo* is the form alone, or, as some say, the union of the form with the matter.

XV. – FIRST CONCLUSION: "The *terminus qui*, or adequated terminus, of generation is neither the form alone nor the matter-form composite, but the supposit."

1st Argument. The *terminus qui* is that to which generation ultimately tends and in which it ultimately rests. But generation ultimately tends to constitute the thing in its integral and perfect *esse*, and it rests only if it obtains complete *esse* that subsists *per se* and in its own right (*sui juris*). Therefore, the *terminus qui* of generation is that in which nature has perfect *esse*, which subsists *per se* and *sui juris*. But a perfect substance, which subsists *sui juris* and *per se*, is a supposit. Therefore, the supposit is the *terminus qui* of generation.

2nd Argument. Since the terminus of generation is *esse*, whatever has being *ut quod* ('as *that which*') is a terminus. But forms and composites do not exist *ut quod*, but *ut quo* ('as *that whereby*'); only the supposit exists *ut quod*. Therefore, the supposit is the *terminus qui*.

XVI. – SECOND CONCLUSION: "The physical composite is the *terminus quo* of the whole, but the substantial form is the partial *terminus quo*."

1st Argument. The *terminus quo*, or the specifying terminus, is that whereby the generated substance is constituted in its own complete species. But the generated substance is constituted in its own species through the substantial form as through a partially distinctive principle and through the whole nature as through a total principle, for the essence and species of material things is not the form alone, but the matter-form composite. Therefore, the substantial form is the partial specifying principle. But the whole nature, or the composite, is the total terminus.

2nd Argument. The *terminus quo* is that by means of which the *terminus qui* is constituted. But the *terminus qui* is constituted immediately through the whole nature and mediately through the substantial form; of course, the supposit immediately includes the nature and it adds onto it many things, and, because nature implies form, the suppositum is constituted by means of the substantial form. Therefore, nature and form are the terminus of generation, the former as the whole, the latter as a part.

XVII. – Third Conclusion: "Properties are the secondary terminus of generation."

A secondary terminus of generation is that which follows from the primary terminus in virtue of the very action that the primary terminus produces. But properties follow in virtue of the action that the primary terminus produces, for given an essence or nature, its properties, which emanate necessarily from the essence, follow from its action. Therefore, properties are the secondary terminus of generation.

Now, recall that properties are not caused by the essence in the manner of an efficient cause (*per modum efficientiae*), but in the manner of a result or of emanation (*per modum resultantiae vel emanationis*). But this emanation can be seen in two ways: taken actively, it is the essence itself as bringing the action of the generating cause to the property itself; taken passively, it is the property as resulting from the essence in virtue of the generating action. This is the common view of the Thomists.

FIFTH ARTICLE

The history of hylemorphism is briefly reviewed

I. – Matter and Form Among the Ancients. To our exposition and defense of the hylemorphic system we should add a few things about its history. The first sketches of this theory are found in Socrates and Plato. Socrates introduced the distinction between essence, which is necessary and the object of science, and the individual, which is contingent and the object of experience; here we have a certain foreshadowing of matter and form: *form* constitutes *essence*, and *individuation* originates from *matter*. Plato admitted the existence of a certain sort of matter, which he claimed was the uncreated, infinite, passive principle of generation, as well as the existence of separate forms, which exists outside of singulars and nonetheless are participated by the singulars. Nonetheless, Plato's view on the true notion and power of matter and form is uncertain and obscure. Aristotle emended Plato's view, made it more moderate, and scrutinized the true nature of matter and form, such that hylemorphism is said to be, and truly is, an Aristotelian system.

The Stoics claimed there were two things in the natures of things from which all things are made: *matter* and *cause*. Matter is inert, passive, indeterminate; cause forms and determines matter. In this respect it is similar to substantial form; but it differs from it in many respects, for it does not give first *esse*, but rather is conceived by the Stoics as being a complete and infinite substance that is in all respects one.

The opinion of the Neo-Platonists approximates that of Plato and Aristotle. They conceive of prime matter as a general substrate of change that is unformed, indeterminate, destitute of any good whatsoever; but form as a simple, indivisible, substantial power. They also seem to attribute living form to all things and to fall into hylezoism. Hylemorphism is transmitted to the Arabs, namely, Avicenna, Averroes, Avicebron, etc., even though they do not agree with Aristotle in all things.

II. – Matter and Form in Christian Philosophy. The Fathers of the Church, who were almost entirely concerned with the things of the faith, cared little about the philosophical constitution of bodies.

Nonetheless, mention is made of matter and form in Origen[74] and in Gregory Nazianzen: "The Deity is above *matter* and *form*, in which bodies consist."[75] But especially Augustine makes much of this theory and considers it to be a *divine dogma*.[76]

In the Thirteenth Century hylemorphism becomes widespread: William of Auxerre uses it in his treatise on the Sacraments, which he says consist, by a certain analogy, *of matter and form*. It entered the Schools through the works of St. Albert the Great, St. Thomas Aquinas, St. Bonaventure, Scotus, and of others, to such a degree that it seems to be confused with Scholasticism itself. It receives great authority from the Council of Vienna, which defined that the intellectual soul is the form of the human body.

What the Council of Trent teaches on the substance of bread and wine in the Blessed Sacrament of the Eucharist seems also to rest on the Scholastic doctrine on matter and form.

III. – Modern Hylemorphism. From the time of Descartes, as Scholasticism was gradually relegated to the religious orders, hylemorphism was almost banished from modern philosophy; and at the turn of the Twentieth Century it was reputed as obsolete, or even as absurd. But through the effort and works of the neo-Scholastics, Sanseverino, Liberatore, Zigliara, Cornoldi, Pesch, etc., in response to the exhortations of Pius IX and later of Leo XIII, hylemorphism was revived and again entered into ecclesiastical schools, and now it abides there in peace. Many authors, such as Zigliara, Pesch, Farges, De Vorges, Mielle, Nys, and many others, have worked assiduously on this study and have produced excellent treatments of this question.

IV. – The Scholastic System is Substantially Retained Today, but the erroneous applications of previous authors—which were due to their imperfect science of physical things—have been rejected. The rejected applications can be reduced to a few headings. (1) Sublunary bodies are essentially different from celestial bodies. (2)

[74] *Periarchon*, 2.1.

[75] *Orat* 23: "Deitatem esse supra *materiam* et *formam* ex quibus constat corpora."

[76] We cited the testimony of AUGUSTINE above, a. 1, n. 10.

There are four elements in sublunary bodies: fire, air, earth, and water. (3) There are four prime qualities: humid, dry, hot, and cold. In every element there is a combination of two of these: in fire there are the qualities of hot and dry; in air, hot and humid; in earth, cold and dry; in water, cold and humid. (4) The cause that forces the elements to combine is the firmament of stars... (5) Elements, according to their various natures, are carried upward or downward to the place that is convenient to their natures: fire and air ascend to some lofty terminus, and soil and water descend toward the earth, which is the unmoving center of all things.[77]

These errors do not touch the substance of the system. As far as its main headings, hylemorphism can be said to be a certain doctrine (*doctrina certa*), namely: (1) Beyond matter and motion, there is in bodies a formal and dynamic principle, abstracting from its nature. (2) From the union of these two principles there results something that is *per se* one. (3) Matter is indestructible: "nothing is lost." (4) Form is not created, but arises from the subject: "nothing is created."

As to what pertains to substantial change, it can be said that it is a certain doctrine (*doctrina certa*) with respect to the human composite, due to the definition of the Council of Vienna; and even with respect to living things, because it is certain that there is an essential difference between a corpse and a living thing; but with respect to inorganic things, even if there is not full certainty, it is supported with convincing arguments.

Further, it is not at odds with any certain fact from the physical sciences; indeed, it harmonizes perfectly with the constant laws that are observed in chemical compounds and with the phenomena of crystallization, and even with the most recent experiments concerning ions.[78] It also saves the truth that is found in other opinions: it concedes to the atomists that there is matter and motion in bodies; with the dynamists it professes that bodies are ruled by dynamic principles. But it avoids the difficulties of each system. First of all, it explains the apparent antinomies to which bodies are subject. It is completely consonant with the Christian faith, especially in those points that pertain to the union of the soul with the body, the Most Holy Sacrament of the

[77] Cf. P. PESCH, *Philosophia nat.*, n. 291.

[78] See above, q. 1, a. 4.

Eucharist, and the human nature of Christ, matters dealt with by theologians.

There are three main ecclesiastical documents that concern hylemorphism, but they shall be more fittingly considered elsewhere, when we discuss the union of the soul and the body.[79]

Let it be concluded, with Suárez, that: "This dogma (of matter and form) is received in philosophy in such a way that it cannot be denied without great ignorance; and it is so consonant with the Christian faith that its certitude is thereby greatly augmented."[80]

It is also appropriate to write the following testimony of the most learned P. Duhem: "Little by little, however, and as a result of development, mechanist hypotheses encountered obstacles from all sides, which were more and more numerous, and more and more difficult to surmount. Therefore atomistic, Cartesian, or Newtonian systems lost the favor with physicists and made way for methods analogous to those which Aristotle defended. Current physics tends to take up again a peripatetic form."[81]

---

[79] *Phil. nat.* II. P., Tract. 2, q. 2, a. 2.

[80] *Metaphys.* IV *Sent. dist.* 12, q. 1, a. 2.

[81] P. DUHEM, *Le Mixte*, p. 200: "Peu à peu cependant, et par l'effet même de ce développement, les hypothèses mécanistes se heurtent de toutes parts à des obstacles de plus en plus nombreux, de plus en plus difficiles à surmonter. Alors la faveur des physiciens se détache des systèmes atomistiques, cartésiens ou newtoniens pour revenir à des méthodes analogues à celles que prônait Aristote. La physique actuelle tend a reprendre une forme péripatéticienne." For this reason, among the 24 Thomistic Theses approved by the Sacred Congregation of Studies on 27 July, 1914, the following are included. Thesis no. 8: "Corporeal creatures are, as far as their very essence, composed of potency and act; which potency and act are, in the order of essence, given the names 'matter' and 'form'" (*VIII. Creatura corporalis est quoad ipsam essentiam composita potentia et actu; quae potentia et actus ordinis essentiae, materiae et formae nominibus designantur*). Thesis no. 9: "Neither of these parts [i.e., matter or form] has *esse per se*, or is produced or corrupted *per se*, or is placed in any one of the ten categories, except reductively, as a substantial principle" (*IX. Earum partium neutra per se esse habet, nec per se producitur vel corrumpitur, nec ponitur in praedicamento nisi reductive ut principium substantiale*).

THIRD QUESTION

On the properties of bodies

Having resolved the question on the essential principles of bodies, both in general and in particular, a consideration on the properties that are concomitant to the essence of bodies now follows. Now, the first accident that is immediately received in a corporeal substance, and which is the proximate subject of accidents, is quantity. Hence, we must discuss quantity and those things which result from it; and then qualitative properties, which are received in the substance by means of quantity.

FIRST ARTICLE

On quantity itself[82]

I. – Definition of Quantity. Because concrete things are better known than abstract things, Aristotle defines quantity by giving a definition of the *quantum* (a quantified thing; lit., a 'how much'): "'*Quantum*' is said of that which is divisible into those [parts] that are within it, of which each or every one is naturally something that is one and something that is a 'this'." (*Dicitur autem quantum quod est divisibile in ea quae insunt, quorum utrumque aut singulum unum aliquid et hoc aliquid natum est esse.*)[83] It is said to be:

---

[82] On quantity and those things which are obtained from it, namely, the continuum, etc., cf. ARISTOTLE and ST. THOMAS, *Metaphys.* V; JOHN OF ST. THOMAS, ALAMANNNUS, CONIMBRICENSES, COMPLUTENSES, GOUDIN, GUERINOIS, MAILHAT, etc. in their *Logics* and *Physics*; PESCH, *Phil. nat.*; SAN SEVERINO, *Cosmol.*; DE SAN, *Cosmol.*; DE MARIA, *Cosmol.*; DE VORGES, *Anal. De Phil. Chrét.*, *La catégorie de la quantité*, Feb. 1888; MIELLE, *De Substantia corporalis vi et ratione*; FARGES, *L'Idée de continu dans l'espace et le temps*; MIELLE, *Revue Thomiste* tom. V, p. 763; NYS, *Cosmol.* nn. 153ff.

[83] *Metaphysics* 5.13. Cf. ST. THOMAS, *In V Metaph.*, lect. 15. Here Hugon reviews three very important physical (and metaphysical) notions. First and foremost, there is the notion of the *quantum*, which means a substance that possesses quantity as one of its attributes (just as something is said to be a good, a *bonum*, because it possesses goodness or *bonitas*, so something is said to be a *quantum* because it possesses *quantitas*). But in order to define the *quantum*,

(1) *Divisible.* In the *quantum* are included divisibility, plurality, and parts. But parts can be *physical*, such as matter and form with respect to a body; or they can be *potential* (*potestativae*), such as the intellect and the will with respect to the soul; or they can be *subjective*, such as 'man' and 'horse' with respect to 'animal'; or they can be *integral*, which are composed in such a way that, after being divided, each remains as something that is one (*aliquid unum*); as, when a piece of wood is divided into many fragments, each of the parts retains a certain unity. Further, the divisibility of the *quantum* is not that of physical, or potential, or subjective parts, but only that of integral parts. Hence, the plurality of the *quantum* is sensible and pertains to position or number.

(2) *Into those [parts] that are within it.* The meaning is this: a *quantum* is divisible into the parts that are within the *quantum* in act. This is said in contrast with the division of mixtures. The mixture, indeed, is divisible into other things, namely, into elements; but elements do not persist formally in act in the mixture, but only virtually.

(3) *Each or every* [*part*] *is naturally something that is one and something that is a 'this'*; that is to say, it is natural for each part to be something that is one (*unum aliquid*); and something that is a 'this' (*hoc aliquid*) is [something that] can be pointed to (*demonstrabile*), even after the division has been made. This is said in order to rule out division into essential parts, which are matter and form. Indeed, after the division into matter and form has been made, neither matter nor form remain as one something that can be pointed to, because after the division neither remains existing *per se*; but, however, when a division into integral parts

---

Hugon—via Aristotle—speaks of the related notions of *hoc aliquid* and *unum aliquid.* A *hoc aliquid*, literally, a 'this something', is simply a particular thing, a primary substance, anything that is a concrete substance, and not just an abstraction or generalization of the mind—a mere species, or secondary substance. Thus, this dog next to me is a *hoc aliquid*, a particular substance, whereas the notion of 'dog' in general is not a *hoc aliquid*, because it is not a particular, concrete thing in the world, but a species, the mind's generalization of what all particular dogs share in common. (Cf. ARISTOTLE, *Categories* 5.) An *unum aliquid* is something that is naturally one, a unitary substance, such as a dog, as opposed to what in modern terminology can be called an aggregate of substances, such as a pack of dogs, which is merely a collection of substances to which we assign a collective name. – ***The Translator.***

has been made, each of the parts persists as something that is one and that can be pointed to.

II. – QUESTIONS. We already had a discussion in logic concerning the division of quantity into continuous and discrete, but the present inquiry concerns continuous and permanent quantity. Now, the Idealists and Kantians contend that quantity does not enjoy any objective reality, but is a mere affection of the perceiving subject. But the Nominalists and Cartesians claim that quantity is not essentially distinct from the essence of a body, and so that a corporeal substance is extended by its own power. The Scholastics are divided concerning the proper constitutive principle of quantity. We ask, therefore, whether quantity possesses objectivity, whether it is an accident that is really distinct from the substance, and what is its proper and formal *ratio*.

III. – FIRST CONCLUSION: "Quantity exhibits a true objectivity and reality."

Proof. Sensation is caused by external bodies. But external bodies can only act upon the senses if they depend on quantity. Therefore, sensation is caused dependently on quantity. Therefore, the reality of sensation depends on the reality of quantity; hence, unless quantity is real, sensation will be vain and illusory. ***Proof of the Major.*** If sensation is not caused by external bodies, it must be produced either by God, or by some other genius who deceives us, or by the soul itself. But it is clear that God is not the only cause or the proximate cause of sensation, and that God does not allow us to be deceived perpetually by some genius, and finally that the soul is not the cause of these phenomena, whether freely or necessarily; for the soul experiences the aforesaid phenomena even against its will; and often these phenomena do not occur, even though all those things that are required on the part of the soul are present. Therefore, the only alternative that remains is that sensation originates from things. ***Proof of the Minor.*** Bodies cannot act upon the senses except in dependence upon sensible accidents. But the proximate subject of sensible accidents is quantity. Therefore, bodies cannot affect the senses except in dependence upon quantity; and consequently, the reality of sensation depends on the objective reality of quantity. ***Confirmation.*** Every activity and motion of bodies presupposes quantity and is founded in quantity, and every contact requires extended quantity. Therefore, if we remove real and objective quantity, we will destroy every activity of the corporeal world, nor would there remain anything except perpetual sterility and inertia. Moreover, the thesis is established from our refutation of dynamism and our proof that there is in bodies a material principle which is the root of extension.

IV. – SECOND CONCLUSION: "Quantity is an accident that is really distinct from corporeal substance." This is the common view of the Thomists, Scotists, Suárez, and all the Scholastics.

Proof. 1st Argument. Quantity in general is related to corporeal substance as this quantity in particular is related to this corporeal substance in particular. But this quantity does not belong to the essence of this substance. Therefore, neither does quantity in general belong to the essence of corporeal substance in general. – The comparison employed in the major is *per se* known. ***Proof of the Minor.*** If it were the case that this quantity belongs to the essence of this substance, then it would follow that when something grows in the order of quantity it also would grow in the order of substance. But substance does not increase or decrease, but is considered indivisible. *Therefore.* ***Confirmation.*** Nothing is indifferent to that which is essential to it. But this substance is indifferent to this determinate quantity; the totality of a substance is contained in a great or a small quantity. Therefore, quantity does not belong to the essence of a corporeal substance.

2nd Argument. That which has *esse secundum quid* is really distinct from that which has *esse simpliciter.* But quantity only has *esse secundum quid.* Therefore, it is really distinct from substance, which has *esse simpliciter.* ***Proof of the Minor.*** Quantity signifies being extended in parts; but being extended in parts is *esse secundum quid.* For *esse secundum quid* is that which presupposes some other *esse.* But to be extended in parts presupposes some other *esse*, for a thing's *esse* is prior to a thing's being extended and diffused in space. Therefore, being extended in parts is *esse secundum quid.* ***Confirmation.*** Any composition that follows essential composition is adventitious and accidental. Indeed, in one and the same *ens* there cannot be two essential compositions. But integral composition, which is given by quantity, follows the essential composition of matter and form. Therefore, integral composition is adventitious; therefore, quantity only gives an adventitious *esse.* The reason for the minor is that nothing in a thing is conceived prior to the union of matter and form, nor can any integral composition be prior to the composition that results from the union of matter and form, and which all other composition presupposes. Hence, Aristotle rightly concludes that: "Length, breadth, and depth are certain quantities, but not substance."[84]

3rd Argument. Substance is *intelligible* with respect to the intellect, whereas quantity is a *common sensible* with respect to sense. But those things which are referred to faculties of such diverse orders cannot but be distinct. Therefore, substance and quantity are distinct. The argument can be pressed in this manner: If quantity were identical to substance, it would not be sensible, like substance itself, which is known only to the intellect. But, if quantity were not at all sensible, then all sensation would disappear. Therefore, unless quantity and substance were really distinct, sensation would be empty and would disappear. ***Proof of the Minor.*** In order for sensation to be elicited, extension must move the senses. But the senses can only be affected if something sensible exists. Therefore, if quantity is not sensible, sensation is destroyed. Therefore, those who reject the distinction between quantity and substance are prone to fall into idealism.

4th Argument. Something that is immediately receptive of accidents is really distinct from an intermediary subject, just as an immediately operative principle is distinct from a principle that is only mediately operative. But quantity sustains accidents immediately, whereas corporeal substance does so only mediately. Therefore, quantity really differs from corporeal substance. ***The truth of the major*** is evident from the established comparison. ***Proof of the Minor.*** A corporeal substance sustains accidents by reason of its matter in the same way that it acts by reason of its form. But it does not act immediately by reason of its form, but through quality, as we have shown throughout. Therefore, it does not sustain accidents by reason of its matter immediately, but through quantity. These are arguments from reason that establish the thesis. The proofs of philosophy certainly make a strong case for it; but theologically the question is resolved more easily. If, of two things, one ceases to be and the other remains, these things cannot be identified, just as things that can be separated are really distinct. But in the Most Holy Sacrament of the Eucharist there remains the quantity of bread and the quantity of wine, while the substance no longer remains. Therefore, quantity and substance are really distinct.

---

[84] *Metaphysics* 8.

But this argument pertains to the theologians. The issue of how accidents can remain without a substance will be discussed in *Metaphysics: Ontology*, Treatise III, q. 2.

V. – ANOTHER QUESTION. It is clear, therefore, that substance in itself is not extended in act; it is also certain that corporeal substance of itself necessarily consists of quidditative parts, namely, matter and form; but it is disputed whether corporeal substance, of itself and independently from quantity, has integral parts or is altogether indivisible. There are two opinions among the Scholastics. Some think that corporeal substance in itself consists of parts—not parts of extension, but integral parts—which they call *entitative* parts; hence, material substance is divisible independently from quantity. This is the view of Suárez, Fonseca, Babenstuber, Goudin, and, among more recent authors, Lahousse and Mendive.

Others defend the view that corporeal substance of itself lacks integral parts and is indivisible. This is the view of St. Thomas, Capreolus, the Conimbricenses, the Complutenses, John of St. Thomas, Alamannus, and among more recent authors, De San, De Maria, De Vorges, Mielle, and Nys.

VI. – THIRD CONCLUSION: "Corporeal substance, of itself and independently from quantity, lacks integral parts altogether; it is simple and indivisible."

Proof from the Authority of St. Thomas: "Matter is divided into parts only insofar as it is understood as having quantity; if quantity is removed, the substance remains indivisible."[85] And elsewhere he says that, "if quantity is removed, every substance is indivisible."[86]

Proof from Reason. 1st Argument. That from which the definition of a thing abstracts does not belong to the essence of a thing. But the definition of corporeal substance abstracts from integral parts. Therefore, having integral parts does not belong to the essence of corporeal substance; for this reason, substance, in itself and independently from quantity, does not have integral parts. *Proof of the*

---

[85] *Summa theologiae* Ia, q. 50, a. 2: "Materiam autem dividi in partes non contingit nisi secundum quod intelligitur sub quantitate, qua remota, remanet substantia indivisibilis."

[86] *Summa contra gentiles* 4.65: "Remota quantitate, substantia omnis indivisibilis est." Hugon has *hominis* instead of *omnis*."

Minor. In the definition of substance, three things can be considered: the notion of composite, the notion of matter, and the notion of form. But (1) the notion of composite does not imply integral parts; for although a composite requires a plurality of accidents, nonetheless this requirement is not plurality itself. (2) The notion of matter signifies only the subject of change; but the notion of a subject of change does not imply a multitude of parts. (3) The notion of form signifies an act, and act of itself is simple. Therefore, the definition of substance abstracts entirely from integral parts.

2nd Argument. To have integral parts is to receive quantity. But substance obviously does not have quantity without quantity. Therefore, it is impossible for substance to have integral parts independently from quantity. *Proof of the Antecedent.* It belongs to the notion of integral parts that they are united only at their extremes, and not throughout. But to be joined at their extremes presupposes that they have extremes, points, and the like, which belong to quantity. Therefore, to have integral parts is to receive quantity. *Proof of the Major.* Those things which are united throughout all their parts penetrate each other, like matter and form; but these are no longer integral, but quidditative parts, one of which is with the other and in the other. Therefore, it belongs to the notion of integral parts that they are united at their extremes and not throughout.

3rd Argument. If substance of itself had integral parts, then the quantity that comes to it would lack its primary effect. For, indeed, if substance were colored according to its essence, then color would not be able to give it its primary effect; therefore, for the same reason, if substance were already extended according to its essence, extension would lack its primary effect, which is to extend the substance in parts.

VII. – Difficulties Resolved. ***1st Objection.*** If corporeal substance does not in itself have parts, it would not in itself be distinct from spiritual substance. ***Reply.*** *I deny the conclusion.* Although substance before quantity does not consist of parts, it has nonetheless the capacity and the requirement of having them; in this respect it is sufficiently distinct from spiritual substance, which neither requires, nor is capable of, parts.

2nd Objection. If God were to conserve a substance in existence without quantity, then the substance would remain either with extended parts or without them. If the former, then the thesis is false; if the latter, then it would follow that all parts would be reduced to a single point, and thus for a time there will remain parts in a point without quantity. Therefore, a substance without quantity has parts. ***Reply***. Under that hypothesis [i.e., that God conserved a substance without quantity], neither the substance would remain with extended parts nor would the parts be reduced to a single point; rather, that substance would become something indivisible outside of the genus of quantity. "Hence, in that substance," says John of St. Thomas, "there would not be any motion or physical place; rather, it would be in the universe as a part of the universe, and not as located in a place: all these imaginings are to be removed, because they follow quantity as located in place, as Cajetan rightly explains (*Summa theol.* Ia, q. 52, a. 1, towards the middle). Hence, that substance is neither distant, nor positively anywhere; but only has its existence without a place, as a thing outside the world, or as an angel that does not act."[87]

VIII. – FOURTH CONCLUSION: "The proper and essential *ratio* of quantity is that of an extension of parts that is ordered to the whole, or to have parts outside of parts, that is to say, parts of which one is not the other and one is outside of the other."

In order to understand the conclusion it must be noted that five *rationes* can be distinguished in quantity:

The first is the *extension of parts among themselves* (*extensio partium in se*), or in their order to the whole itself, just as it is explained in the very words of the conclusion. This kind of extension is also said to be *internal quantitative extension,* or the 'positional extraposition' (*extrapositio situalis*) of parts in the whole, that is, insofar as one part is outside of another.

The second is the *extension of parts in place*, or their circumscriptive extension, which consists in the fact that one part is outside of the other, not only with respect to their entity, but also with respect to their place, such that one is outside of the place of the other, and one is in one part of place and another in another part.

[87] *Log.* IIa P., q. 16, a. 1, *Solv. arg.*: "Unde non esset in illa substantia aliquis motus sicut neque locus physicus, sed solum esset in universo tamquam pars illius, non ut locatum in loco: omnes istae imaginationes tollendae sunt, quia sequuntur quantitatem ut locatam, ut bene docuit Cajetanus (I P. Q. 52, art. I, circa medium)."

The third is *impenetrability*, or the impossibility for one locally extended body (having quantity in act) to occupy the same place with another body that is also locally extended.

The fourth is *divisibility*, or quantity's ability to make united parts be separated and remain nonetheless as something that is one (*aliquid unum*) after the division has been made.

The fifth is *measurability*, whereby we are able to determine the magnitude or smallness of things. This *ratio* of measure can be found both in discrete and in continuous quantity. Hence, things that have *oneness*—which is the principle of number—have measure; number is defined as "a multitude measured through oneness." And this is the proper *ratio* of measure. But, since the *minimum* is a certain imitation of oneness, the minimum also has the *ratio* of measure. Hence, measure is unanimously defined as, "That whereby the smallness or magnitude of a thing is known."

"Note that quantity is an *intrinsic* measure of the subject within which it is, insofar as it is useful for knowing its magnitude; and also an *extrinsic* measure with respect to other things, insofar as we use it to measure other things; for example, we use a fathom to measure a piece of cloth and its quantity."[88]

Certain philosophers thought, with Simplicius, that the *ratio* of quantity consists in measurability; others in divisibility; yet others, such as Arriaga and Oviedo, in impenetrability. The more common opinion is that which expresses our conclusion, and which the Neo-Scholastics defend.[89]

Proof. 1st Argument (from the exclusion other opinions).

(1) Measurability is not the formal *ratio* of quantity. Measurability presupposes plurality and the measurable presupposes extension. But that which presupposes something prior is not a formal constitutive principle of that prior thing. *Therefore.*

(2) Nor is divisibility the formal *ratio* of quantity. For divisibility denotes the separability of parts. But being extended in parts is prior to being separated into parts. Therefore, divisibility requires something prior, namely, extension in parts.

---

[88] P. MAILHAT, O.P., *Log.* P. I. disp. 6: "Adverte quantitatem esse mensuram *intrinsecam* subjecti cui inest, quatenus deservit ad dignoscendam ejus magnitudinem; et *extrinsecam* respectu aliam rerum, quatenus ea utimur ad res alias mensurandas, puta, ulna ad mensurandum pannum ejusque quantitatem."

[89] Cf. Nys, *Cosmologie*, nn. 177-186.

(3) Nor is it impenetrability. For, indeed, having parts outside of parts is prior to excluding parts from being in the same place; this exclusion results from the distinction of parts.

(4) Nor is it extension ordered to a place. Just as *esse simpliciter* is prior to *esse tale* (being such and such), so there being parts is prior to there being parts in a determinate place. Therefore, extension ordered to a place presupposes something. It remains, therefore, that the formal constitutive principle of quantity is the extension of parts in the whole.

2nd Argument (from the very notion of formal constitutive principle). The formal constitutive principle of a thing is that which is first in that thing and which is the root of all that is attributed to it. But the extension of parts ordered to the whole is first; for measurability, divisibility, impenetrability, and extension in place presuppose parts placed outside of parts being ordered to each other, as is clear from the previous argument. – It is the root of the rest. For, due to the fact that quantity has parts outside of parts, it follows:

- that one part naturally must be outside of the place of the other, and consequently that one part exclude another from the same place: hence extension and impenetrability;
- that parts can be separated, hence divisibility,
- and finally that quantity be measuring and measurable.

Therefore, the extension of parts in the whole is the formal principle of quantity.

IX. – COROLLARIES. The essence of a thing cannot be separated from that thing; hence, it is impossible for quantity to exist without internal extension; but it is not impossible at least for secondary properties to be separated from the essence by God. Hence, it is possible for there to be quantity without local extension, without impenetrability, etc. In this way, in the Eucharist, the quantity of the body of Christ is present as far as the formal constitutive principle of quantity, namely, internal extension, but not as far as local extension. In other words, in the body of Christ, each part is distinct from the rest: the head is not the neck, and the neck is not the chest; but the head, the neck, and all the other parts are present in the same place. Hence, dimensional quantity is there in the mode of substance, which is entirely in the whole and entirely in each part.

Thus, when bodies miraculously penetrate each other, quantity is not destroyed, for indeed parts do remain outside of parts, but only a secondary effect of quantity is impeded, namely, that one part expels another from the same place.

X. – Difficulties Resolved.

*1st **Objection***. A definition expresses the essence of something. But quantity is defined as, "that which is divisible...." Therefore, divisibility is the essence of quantity. ***Reply***. *I distinguish the major*: that an *essential* definition expresses the essence of the thing, I concede; but that a *descriptive* definition expresses the essence of the thing, I deny, for it describes the thing only through its properties. *I contradistinguish the minor*: that the *descriptive* definition of quantity is "that which is divisible...," I concede, for Aristotle intended to describe quantity on the basis of something that is better known to us (*quoad nos*), namely, on the basis of divisibility; but that this is the *essential* definition of quantity, I deny, for quantity is properly and essentially defined as: "An accident that extends substances into parts."

*2nd **Objection***. Quantity denotes the dimensions of length, breadth, and depth. But having three dimensions requires external extension. Therefore, quantity denotes external extension. ***Reply***. *I distinguish the major*: that quantity, taken as including with all its effects, both primary *and secondary*, denotes three dimensions, I concede. That quantity, taken as including *only its primary and formal effect*, denotes three dimensions, *I subdistinguish*. That quantity, taken as including only its primary and formal effect, denotes three dimensions which are *ordered to parts*, I concede. But that quantity, taken as including only its primary and formal effect, denotes three dimensions which are *ordered to place*, I deny. *I contradistinguish the minor, and I distinguish the conclusion*: that, therefore, quantity, taken as including all its effects, denotes external extension, I concede; but that, therefore, quantity, taken as including only its primary and formal effect, denotes external extension, I deny.

XI. – It Belongs to Quantity Not to Have a Contrary, Not to Receive More or Less, and to Be the Foundation of the Equality and Inequality of Things.

[*It does not have a contrary.*] Contraries expel each other from the same subject. But quantity does not expel, but rather requires, quantity; for a body requires a surface, and surfaces require lines.

It does not receive more or less according to its essence. For the same quantity is not more a quantity and then less a quantity; nor does a greater quantity, e.g., a line that is a thousand kilometers long, participate more in the *ratio* of quantity than a smaller quantity, e.g., a line of millimeter. And yet, the same quantity can be greater, then smaller, and one quantity can be greater or smaller than another; for this does not require the increase or decrease of the nature, but the addition or subtraction of parts suffices.

It is the foundation of equality and inequality. For just as the *identical things* and *diverse things* are those which concur or do not concur in *substance*, so *similar things* and *dissimilar things* are those which concur or do not concur in *quality*; in the same way, *equal things* and *unequal things* are properly said of those which concur or do not concur in *quantity*.[90]

---

[90] There are many other aspects of quantity that pertain to the natural and mathematical sciences which, for that reason, the philosopher does not study. We also relegate to scientists those things that concern molar quantity (mass). This consideration, of course, is not directly philosophical. Concerning this matter one may consult NYS, *Cosmologie*, nn. 102ff. This doctrine which we have presented is summarized in the following proposition approved by the Sacred Congregation of Studies, 27 July, 1914. Thesis 10: "Although corporeal nature implies extension into integral parts, it is not the same for a body to be a substance as for a body to be a *quantum*. Indeed, a substance is, by its own *ratio*, indivisible; not, of course, in the manner of a point, but in the manner of that which is outside of the order of dimension. But quantity, which gives extension to substance, is really different from substance, and is an accident in the true sense of the term." (*X. Etsi corpoream naturam extensio in partes integrales consequitur, non tamen idem est corpori esse substantiam et esse quantum. Substantia quippe ratione sui indivisibilis est, non quidem ad modum puncti, sed ad modum eius quod est extra ordinem dimensionis. Quantitas vero,quae extensionem substantiae tribuit, a substantia realiter differt, et est veri nominis accidens.*)

SECOND ARTICLE

On the continuum

I. – The Notion of the Continuum. Having discussed the notion of quantity, now we discuss the concept of the *continuum*, which arises from the *ratio* of quantity.

Now, the continuum is said of "that whose extremes are one"; but the contiguous is said of "that whose extremes are together." For things to be contiguous it suffices that their parts be juxtaposed, e.g., such as two joined surfaces; but in order to constitute a continuum, simple juxtaposition is not sufficient; parts must be joined by a common terminus and that they form something that is naturally one. *Discontinuous things* are those whose parts are separated by intervals; *distant things* are those between which there is nothing intermediary that belongs to them.

II. – Division of the Continuum. The Scholastics make a fivefold distinction of the continuum: *lines*, which are divisible lengthwise; *surfaces*, which are divisible on their side; *bodies* or *solids*, which are divisible throughout; *motion*, which is divisible into the part prior to the acquisition of the terminus and the part posterior to the acquisition of the terminus; and *time*, which is divisible into the part prior to the motion and the part posterior to the motion. There is also a distinction between the *perfect* continuum, in which all the parts are together without any void in between, and the *imperfect* continuum, in which, beyond all the parts that are full and perfectly together, there are other parts that are separated by certain intervals, as in sponges. Nor should one ignore the division of the continuum into *homogeneous* continuum and *heterogeneous* continuum. The former occurs when all the parts in the whole are of the same *ratio*, as in the case of water, all of whose parts are water; and the latter occurs when the diverse parts of the whole are not of the same nature, but of different natures; thus, not all of man's parts are flesh or bones.

III. – Diverse Opinions Concerning the Constitution of the Continuum. The opinion of many thinkers was that the continuum was made up only of indivisible things. We here refute three opinions. The first is that of the atomists, who claim that the continuum is formed of indivisible atoms, or rather they deny that there is such a thing as a true continuum, since they assert that atoms are absolutely separated by empty intervals. The second is that of the dynamists, who think that the continuum consists of simple substances or of inflated points. The third is that of Zeno and of the ancients who claimed that the continuum consists of indivisible points.

IV. – First Conclusion: "The Continuum does not consist of indivisible parts alone, but of both continuous and indivisible parts."

In our arguments against the atomists we have shown that atoms are not separated all the way around by empty intervals, but that there is a true continuum within atoms themselves, within the ether that lies between them, and in substances themselves. We also refuted in the same place the arguments that our adversaries offer from porosity, from rarefaction and condensation, and from vibratory motion. We admit a certain discontinuity, and the scientific facts do not allow us to posit an absolutely perfect continuum as the ancients posited. And yet, reason shows that there is no action at a distance and, therefore, that atoms and molecules must touch each other on all sides. This is sufficient for the *ratio* of the continuum.

Now, we already sufficiently discussed monads and inflated points. Hence, we now refute the third opinion specifically.

Proof of the Conclusion. The parts of the continuum must be such that adding one to another will result in something greater in extension. But adding something indivisible to something indivisible does not result in something greater in extension. Therefore, the continuum cannot consist only of indivisible parts. ***The major is evident***, for experience attests to the fact that the continuum grows and becomes more extended through the addition of parts. ***Proof of the Minor***. Indivisible things, because they do not have extremes or middle parts, touch each other entirely. But those things which touch each other entirely—which come within each other and totally penetrate each other—do not make up something greater in extension. Therefore, adding something indivisible to something indivisible will not result in something greater in extension.

The Scholastics, therefore, rightly infer that the continuum is composed of continuous and indivisible parts: thus a line is made up of partial lines; a surface, of partial surfaces; a body, of parts that have three dimensions; motion, of partial motions; time, of partial successive durations.

V. – SECOND CONCLUSION: **"In the continuum, nonetheless, there are truly indivisible parts, some of which are continuative of other parts and others which are terminative of the extremes."** This is the common opinion of the Thomists.

Our opinion holds a middle ground between two opposite opinions, namely, between that which asserts that the continuum consists only of indivisible things, and that which denies that there are any indivisible parts in the continuum.

Proof of the First Part, namely, that there are indivisible parts that are continuative of other parts. This is the difference between the continuum and the contiguous: that, in the contiguous, parts are only juxtaposed; but, in the continuum, parts are joined by common termini. But common termini that join parts must be indivisible. Therefore, in the continuum there are indivisible parts that are continuative of other parts. ***Proof of the Minor***. If the terminus that joins parts is not indivisible, then it has parts, and so its parts need to be joined by some other intermediary nexus, which itself has parts that are themselves to be joined by yet another intermediary nexus; and thus we have an infinite regress.

Proof of the Second Part, namely, that there are indivisible parts that are terminative of the extremes. Bodies must have a real and physical contact without penetrating each other. But, physical contact among bodies without mutual penetration is impossible unless the parts that are terminative of the extremes are indivisible. *Therefore*. ***Proof of the Minor***. If the parts that are terminative of the extremes are not indivisible, then this would only denote either a negation of further extension, or those parts themselves as connoting such negation. But this negation is obviously not sufficient for physical contact, because in that case bodies would touch each other in nothingness; nor is it sufficient for there to be parts that connote negation, because they would be divisible parts, and the divisible, if it is joined to the entire divisibility of another, does not touch by remaining outside of the other part, but by entering into the other part, which is precisely what mutual penetration is. Therefore, if the parts that are terminative of the extremes are not indivisible, it is impossible for physical contact to occur without mutual penetration.

— These indivisible parts are five in number: *the point*, which lacks all divisibility and extension; *the line*, which lacks divisibility on its side; *the surface*, which lacks divisibility within; *being moved*, which is not divisible into prior and posterior; and *the instant*, which is not divided into before and after, but is a simultaneous whole.

VI. – THIRD CONCLUSION: "These indivisible parts, which are continuative or terminative of other parts, are not mere beings of reason (*entia rationis*), but are modally distinct from parts, and reductively (*reductive*) pertain to the category of quantity."

Proof of the First Part. The positive nexus and the positive terminus of real beings are not mere beings of reason (*entia rationis*). But an indivisible part that is continuative of others is a positive nexus of real parts, and an indivisible part that is terminative of others is a positive terminus of real parts. Therefore, these are not merely beings of reason.

Proof of the Second Part. The indivisible and the divisible are not the same. But there are divisible parts. Therefore, indivisible parts are distinct from divisible parts.

– But a modal distinction is sufficient, nor is it possible to make a real entitative distinction.

For those things which are really entitatively distinct can be separated, that is, one can exist outside of the other. But it is impossible for the nexus of parts to exist outside of the parts themselves that it joins or for the terminus to exist outside of the parts of which it is the terminus. Therefore, the indivisible nexus and terminus of parts are not really entitatively distinct from the parts.

Explanation of the Third Part. These indivisible parts are not all properly and *per se* in the category of quantity, because not all enjoy extension; but reductively, at least, they are referred to the category of quantity, for modes belong to the category of 'thing'.

VII. – WHETHER THE CONTINUUM IS DIVISIBLE IN ALL ITS PARTS. The parts of the continuum belong to two genera: some are signate and determinate, and are called *partes aliquotae*[91]; but others are indeterminate and are called proportional parts (*proportional parts*).

[91] *Partes aliquotae* (literally, 'several parts') are repeated parts, so-called because they are repeated several times within the whole. – ***The Translator.***

Determinate parts have a certain extension, such that when repeated several times they equal to and produce the whole. Centimeters belong to this kind of part with respect to meters because, repeated a hundred times, they equal to and produce a meter. But indeterminate or proportional parts are the halves (*medietates*) of the continuum, and the halves of the halves, and so on *ad infinitum* under the abstract *ratio* of 'half' (*sub praecisa ratione medietatum*), abstracting from determinate greatness or smallness.

Having noted these things, we now respond to the question: The continuum can be divided in all its *partes aliquotae*, or determinate parts, but not at all in its proportional parts. For determinate parts are equal to the continuum insofar as they educe the continuum when repeated several times; hence one can reach the last one. But if one can reach the last one, then one can carry out the division until the last one. Therefore, the continuum can be divided into all of its *partes aliquotae*.

But the proportional parts are such that, after any division they can always be further divided in half, and those halves in half, and so on *ad infinitum*. Since, therefore, one can never reach the last one, it is impossible for the continuum to be resolved into all its proportional parts.

VIII. – Whether the Continuum is Divisible *ad infinitum*. There are three opinions on this matter. The Cartesians, with whom the Atomists agree, believe that quantity is divisible *simpliciter* (*simpliciter divisibile*) *ad infinitum*. The dynamists believe that divisibility *ad infinitum* is absolutely impossible. The common doctrine of the Scholastics is expressed thus:

IX. – Conclusion: "Quantity in itself, that is, considered mathematically, is divisible *ad infinitum*; but physically, that is, as it is found in physical things, it cannot be divided *ad infinitum*."

Proof of the First Part. Whatever has an infinite number of proportional parts is divisible *ad infinitum*. But mathematical quantity has an infinite number of proportional parts. Therefore, it is divisible *ad infinitum*. *The major* is evident on its own terms, and *the minor* is clear from the preceding. Proportional parts are halves, and the halves of the halves. But to any half one can always add another half, and so on *ad infinitum*. *Therefore*. ***Another proof.*** As long as something remains extended, it is divisible at least in itself, since it retains separable parts.

But a mathematical quantity, however small it may be, is essentially extended. Therefore, mathematical quantity is always indefinitely divisible. *Proof of the Minor.* A quantity, however small it may be, retains its *ratio* of quantity. But the *ratio* of quantity denotes extension. Therefore, a quantity, however small it may be, is essentially extended and can never be simple. Consequently, it is divisible *ad infinitum.*

Proof of the Second Part. A quantity, taken physically and concretely, is measured according to its *partes aliquotae.* But the *partes aliquotae* are finite, such that at some point one reaches the last one, as was said. Therefore, a quantity, taken physically and concretely, cannot be divided *ad infinitum.*

Further, a continuous body, as it exists concretely, has a determinate form and its own activity. But a form, in order to exist, requires a determinate quantity; unless this quantity is preserved, the thing will cease to be. Also the activity of bodies requires a determinate quantity; if this quantity is removed, the thing will not be able to exercise its own power, but will be broken down through the excess of the other and will be converted into something else. Therefore, the continuum as it exists physically and concretely cannot be divided infinitely and indeterminately.[92]

X. – WHETHER THE PARTS ARE IN THE CONTINUUM IN POTENCY OR IN ACT. There are two opinions. The first claims that parts are actually distinct before any division. This is attributed to Hurtado, Suárez, and Toledo. The other, which is that of Aristotle, St. Thomas and many Scholastics, teaches that before division the parts are only potentially distinct.

These opinions can be reconciled to some degree. For parts can be considered in two ways: (1) under the *ratio* of some reality, and (2) under the formal *ratio* of part. Hence the question is twofold:

---

[92] ST. THOMAS, *In De sensu*, lect. 15: "A *mathematical* body is divisible *ad infinitum,* since in it we consider only the *ratio* of quantity, in which there is nothing that is contrary to infinite division. But a *natural* body, which is considered under a form, cannot be divided *ad infinitum,* because, when it is reduced to a minimum, it is immediately converted into something else due to the weakness of its power" (*Corpus mathematicum est divisibile in infinitum, in quo consideratur sola ratio quantitatis in qua nihil est repugnans divisioni infinitae. Sed corpus naturale, quod consideratur sub tota forma, non potest in infinitum dividi, quia, quando iam ad minimum deducitur, statim propter debilitatem virtutis convertitur in aliud*).

(1) Whether those things into which the continuum can be divided are realities that are contained *in act* in the continuum before the division itself is done. The response is evidently affirmative: when, for example, gold is divided into many [particles], those things into which it is divided do not acquire a new nature, but retain the same nature that they had before the division. Therefore, parts taken in this sense are present in act in the continuum before the division. For this reason Aristotle defines the *quantum* as: That which is divisible into those things that *are present in it*. From this we can infer the fact that the continuum is *divisible in act*.

(2) Whether parts are not only realities that are contained in act in the continuum, but also unities that are actually distinct from each other; from which would follow that the continuum is not only divisible in act, but also *multiple* in act. If the defenders of the first view think this is the case, then they completely depart from Aristotle and St. Thomas. Hence:

XI. – CONCLUSION: "The parts of the continuum are not actually distinct under the formal *ratio* of part, but only potentially."

Proof from the Authority of St. Thomas: "Before division, the parts of some homogeneous continuum do not have *esse* in act, but only in potency."[93] – "No part of a continuum exists except in potency; hence, neither is a part of fire something in act except after being divided."[94]

Proof from Reason. 1st Argument. If parts are actually distinct, they are either infinite or finite. But both of these hypotheses have impossible consequences. *Therefore. Proof of the Minor.* If they are infinite, there will be something actually infinite in the continuum, which is impossible. If they are finite, they can be counted before the division: but this is inconvenient, for number does not precede, but follows, division. Further, if they are said to be finite, the divisibility of the continuum *ad infinitum* is denied, which the adversaries nonetheless admit.

2nd Argument. If parts are actually distinct, the *esse* of one will be actually distinct from the *esse* of another, and hence there will be as

---

93 *In IV Sent.*, d. 10, q. 1, a. 3, qc. 3, ad 1: "Partes autem alicujus homogenei continui ante divisionem non habent esse actu, sed potentia tantum."

94 *In I Sent.*, d. 8, q. 5, a. 3, ad 2: "[N]ullius continui pars est nisi in potentia: unde nec pars ignis est aliquid actu, nisi post divisionem." Here Hugon erroneously cites *De potentia* q. 9, a. 7. – ***The Translator.***

many actual beings in the continuum as there are parts and, therefore, the continuum will not be something that is *per se* one.

XII. – ON THE HETEROGENEOUS CONTINUUM. The heterogeneous continuum has something in common with the homogeneous continuum: in each there is a single form and a single essence. But the difference lies in this: that in the homogeneous continuum the *form relates to the whole and to the parts equally*; hence it follows that the concrete name of substance can be rightly predicated of both the whole and of the parts; the whole body of water is water, and each part of that body of water is water. But in a heterogeneous continuum, since it has dissimilar and irregular parts, *the form is not equally related to the whole and to the parts*; hence it follows that the concrete name is not rightly predicated of the parts, but indirectly and in relation to the whole. For one does not say: "The arm or the foot is man," but: "is a part of man." We shall discuss the heterogeneous continuum in [our philosophical treatises on] Biology and Psychology (IIa-IIae).

THIRD ARTICLE

On place[95]

I. – False Notions of Place. Corporeal substance, by the very fact that it has quantity (*eo ipso quod sit quanta*), naturally requires that it occupy its own place through its own dimensions. Hence our need to consider place now.

Plato thought that place was the matter of bodies itself; others thought that place was the form or figure. These views hardly require a refutation. For place can change without the located thing changing. But matter or form cannot change without the substance changing. But figure and other intrinsic forms are not torn away from the subject in the same way that place is separated from the located thing.

– Kant contended that place is a subjective form of the knower. But this is false. For the terminus of real and objective motion cannot be a mere figment of the subject. And place is the terminus of real and objective motion; for indeed, bodies really move to acquire a new place, and a subjective concept is not sufficient for something to change place. Therefore, place is not a subjective form.

II. – The Aristotelian Definition of Place. The definition that the Philosopher gives is well known: "Place is the first, immovable surface that contains a body."

It is said to be: (1) *a surface*, that is, a concave surface that receives the located thing; (2) *that contains a body*, that is, that surrounds it: thus the concave surface of air that surrounds a house is the place of the house; (3) *the first* surface, that is, the one that immediately surrounds and touches the contained and located body, in contrast to other intermediate surfaces, which do not surround the located body immediately; (4) an *unmovable* surface; for this is the difference between a container and a place, that a vessel is a mobile receptacle, whereas a place is an immovable receptacle. Further, we are speaking of immovability taken formally; for this reason, the place of a ship in a

[95] Here one may consult: Aristotle, *Physics* 4; St. Thomas, *In IV Phys.*; *Quodlibet* VI; the opusculum *De natura loci*; John of St. Thomas, Goudin, and Guerinois in their *Physics*; Suárez, *Disputationes Metaphysicae*. 51; Sylv. Maurus, *Phys.* 4; Farges, *L'Idée de continu*; Lorenzelli, *Philos. de loco*; Nys, *La notion d'espace*, etc.

river is not, properly speaking, the water that flows and moves, but the relationship and position that it has with respect to the whole river. Generally, the surface is said to be formally immovable insofar as it has such a position in the universe and such and such a distance from fixed points, e.g., from the center and from the poles. Hence, whatever changes my body undergoes, if the surface of the body that surrounds me retains the same distance from the center and from the poles of the Earth, I shall always remain in the same place; but as soon as that distance varies, my place will change.

Now, of course, the center and poles of the Earth are not absolutely immovable, but that is not required for the *ratio* of place; for indeed, a surface is considered formally immovable so long as it retains the same position with respect to something that is relatively immovable. If we inquire further, however, we take the nature of place with respect to the Earth and the planets in relation to something more immovable, namely, the Sun; with respect to the Sun in relation to some center that is even more immovable; and finally with respect to the whole universe in relation to some body, or center, that is *conceived* as altogether immovable that contains all places and all located things. And this first container, whatever its reality may be, can be understood to be the farthest extreme of the corporeal world, outside of which no other body is conceived.

III. – To what Category does Place Belong? Place is referred to quantity, but is not a distinct species of quantity. From the category of quantity it takes the nature of a surface and the *ratio* of container, of something that surrounds the located thing. And this surface is not distinct from the surface that is a species of quantity; for the *ratio* of surrounding a new *esse* does not add to quantity, but only applies the concept to the located thing.

IV. – Division of Place. Place is divided into *extrinsic* and *intrinsic* place. Extrinsic place is the surface that intrinsically surrounds, as we already explained, whereas intrinsic place is the very passive presence of the located thing. We also distinguish between *common* place and *proper* place: *common* place is the surface that does not immediately touch the located thing, whereas *proper* place is the first surface that immediately surrounds the body. Place is said to be *natural* if it conserves the located thing, but *violent* if it is the surface of a harmful body.

V. – The Properties of Place. The primary properties of place are: (1) that it contains the located thing; (2) that it conserves the located thing; (3) that it is adequate to the located thing; (4) that it has the directions of up and down; (5) that it is separable from the located thing; (6) that it is the extrinsic terminus of motion, since it is precisely in order to acquire a new place that the thing moves.

VI. – 'Where' (*ubi*). The accident that results in the located thing from the fact that it is subject to an extrinsic place is called *where* (*ubi*). Many authors think that *where* is a certain entity that is distinct and separable from the located thing, but others claim it is only a being of reason (*ens rationis*). Taking a middle course, we shall say that *where* is something intrinsic that connotes something extrinsic, and that it is distinct from the located thing, just as a mode is distinct from a modified thing.

(1) *It is something intrinsic and real*; for a located body, when it acquires a new *where*, is *really and intrinsically* changed, and this change presupposes a real foundation. Therefore, the change that results in the located thing from the acquisition of a new *where* is not a mere figment of reason, but rather something real and intrinsic.

(2) *It* nonetheless *connotes something extrinsic*, for the place from which it acquires such a change and denomination is a measure that is extrinsic to the located thing.

(3) *'Where' is not an entity that is separable from the located thing.* For it is a measuring up (*commesuratio*) of the located body to the extremes of the body that contains it. But a 'measuring up' cannot be separated from the thing that is measured up. It remains, therefore, that it is modally distinct. *Where*, therefore, can be defined as: "An intrinsic mode that is derived from the extrinsic, surrounding place, and from which the denomination of 'located' is derived."[96] – Recall the other things that we discussed in Logic.[97]

---

[96] John of St. Thomas, *Phil. nat.*, q. 16, a. 1: "Modus intrinsecus proveniens a loco extrinsecus circumscribente, ut a quo provenit principaliter denominatio locati."

[97] *Minor Logic*, Treatise I, q. 1, art. 8.

VII. – In How Many Ways Something can be in a Place. Five ways are assigned. The first is through the measuring up of the dimensions of the located body to the dimensions of its place. This way of being in a place implies two things: (1) that the located thing receives dimensions; (2) that it is placed in that place by means of its own dimensions. This way of being in a place is also called the 'circumscriptive' way.

The second way of being in a place is for a corporeal substance to be immediately located in a place, by means, not of its own dimensions, but of those of another, under which the substance is hidden. Thus the body of Christ in the Eucharist is not in a place through its proper dimensions, but by means of the dimensions of the Host, which nonetheless remain without a subject. Evidently, this is a supernatural way of being in a place.

The third way of being in a place is through virtual contact, or through the application of the power (*virtus*) to the place or to the located thing. Angels can be in a place neither through dimensions, because they do not have dimensions, nor immediately through their own substance, which insofar as it is altogether spiritual is in itself abstracted from place; but they make themselves present in place insofar as they apply their power to that place. And, if indeed their power is bound to one place such that it cannot be exercised somewhere else, they are said to be *definitively* in a place.

The fourth way of something being in a place is through its essence, insofar as it informs a located thing. This belongs to informing forms, whether they are substantial forms or accidental forms, but not at all to forms that are perfectly subsisting.

The fifth way of something being in a place is through its essence, insofar as it gives *esse* to a place and to a located thing and conserves this being and infinitely exceeds it. Now, to produce *esse* and to conserve a thing in *esse* belongs to God alone. And, this way of being in a place is called *ubiquity* and *omnipresence,* because God gives and conserves *esse* to one place and to one located thing in such a way that He gives it simultaneously to each and to all, all the while exceeding all things.

VIII. – Which Things Properly Have '*Where*'. Although there are five ways of being in a place, nonetheless only those things have *where* which are in a place circumscriptively.[98] Hence, that alone is *per se* in a place which possesses quantity and dimensions and which can be circumscribed by something extrinsic. And so, forms, whether substantial or accidental, insofar as they are simple, cannot *per se* be subjected to a place, but only *per accidens*, by reason of the whole in which they are present. Thus, if a body were to be conserved in existence without quantity, it would not be properly speaking in a place; in fact, a body that does have quantity, but is not placed in a place through its own dimensions, is not properly speaking in a place, for it would not measure up to the dimensions of place. A body placed outside of the whole universe would not be in a place because it would lack the extension by which it would need to be circumscribed. What, then, are we to say about the last body that is supposed to be at the limit of the universe? According to the doctrine of Aristotle, it is not *per se* in a place, for, since it is the last body, it does not have above itself or outside of itself anything that contains and circumscribes it. It can be said, nonetheless, to be in a place accidentally and in potency "by reason of the parts, insofar as one continues the other, but is not surrounded by it, because it is not separate from it."[99]

IX. – Whether Two Bodies can Occupy the Same Place. Mutual penetration is the filling of the same place by two bodies. Many philosophers held that mutual penetration can occur naturally, and indeed occurs daily, for we see one body being received within another. But Durandus, then Descartes, Locke, and many among the mathematicians thought mutual penetration was impossible and contradictory. The common opinion (*sententia communis*) of theologians, however, states that many bodies can penetrate each other by divine intervention.

X. – Conclusion: "Although bodies are naturally impenetrable, nonetheless God can bring it about (*fieri tamen divinitus potest*) that many bodies occupy the same place."

Proof of the First Part, namely, that mutual penetration is not natural.

---

[98] Cf. *Minor Logic*, Treatise I, q. 1, a. 8, n. 7.

[99] John of St. Thomas *Phil. nat.* q. 16. a. 3: "Ratione partium, quatenus una continuatur alteri, sed non circumdatur ab illa, quia non est separata ab ea."

1st Argument. That is unnatural which is totally opposed to some natural effect. But mutual penetration is opposed to a natural effect of quantity. Therefore, it is not natural. *Proof of the Minor.* Mutual penetration causes two parts of quantity to be present in the same place. But, since the *ratio* of quantity is to have parts outside of parts, it *naturally* follows that one part is outside the place and position of the other. Therefore, mutual penetration is opposed to a natural effect of quantity.

2nd Argument, from the vanity of the reasons that the adversaries give. One body, they say, is received into another. Therefore, bodies penetrate each other. *Reply.* The adversaries seem to confuse compressibility with penetrability. Further, compressibility is the property whereby a body is able to contract itself and occupy less space, whereby it happens that another body can be introduced into a place that was previously occupied by it but which is now abandoned. But here it is not at all the case that two bodies occupy the same space, but rather one body ceases to be in some part of its prior place and then a new body occupies that free part of the previous place of the old body.

Many facts prove that compressibility in the true sense of the word does occur. In order to explain the vibratory motion of the ether through molecules and the undulations of light waves we must admit either action at a distance or the real compressibility of the substance of ether. If ether were completely filled (*omnino plenus*) and incompressible, how could motion take place? But if compressibility is admitted, then everything can be coherently explained; while one part is compressed, another is expanded, and vice versa, and hence pulsations, vibrations, and undulations can result.

The same can be concluded from the notion of elasticity. When two billiard balls collide, they immediately move in opposite directions. And this implies that the balls are deformed, compressed, and then expanded, all in an instant.

Similarly, the liquefying of gaseous bodies could hardly be explained if in their substance they did not possess the property of compression and expansion.[100]

Proof of the Second Part, namely, that God can bring about the mutual penetration of bodies. First, God could remove actual quantity from two bodies, in which case nothing prevents them from being in the

[100] Cf. FARGES, *L'Idée de continu*, 2e p., no. 3.

same place, just as nothing prevents two angels from being together in the same place. In fact, it can even happen that two bodies are in the same place *circumscriptively*.

Proof. 1st Argument. God can separate from a thing that which is only that thing's secondary effect; for then the essence remains intact. But mutual penetration is only a secondary effect of quantity, for its primary effect is to have parts outside of parts, and it secondarily follows that diverse parts exclude each other from the same place. Therefore, God can separate mutual penetration from quantity, and thus two bodies can be together in the same place.

2nd Argument. Mutual penetration is possible if there can remain a distinction of bodies without a distinction of place. But a distinction of bodies can miraculously remain without a distinction of place. Therefore, mutual penetration can occur miraculously. *The major* stands, for two bodies cannot be in the same place due to their diverse dimensions according to place. *Proof of the Minor*. "God, who is the first cause of all things, can conserve an effect in existence without proximate causes; hence, just as he conserves in the Sacrament of the Altar the accidents without their subject, so he can conserve the *distinction* of corporeal matter without the diversity of place."[101]

XI. – Difficulties Resolved.

1st Objection. It is impossible for the capacity of one thing to contain two things. But, if two bodies are in the same place, the capacity of one body contains two bodies. Therefore, it is impossible for two bodies to be present in the same place. ***Reply***. *I concede the major and deny the minor*. We concede that beyond the capacity of one volume, two volumes cannot be contained. Further, in that hypothesis there are not two volumes in the capacity of one volume, but rather, the two bodies have a common volume and common dimensions.

2nd Objection. If bodies penetrate each other, one part is *with* the other. And the essence of quantity requires that one part be *outside* of another. Therefore, the essence of quantity prevents mutual penetration. ***Reply***. *I distinguish the major*. That one part would be *with* another in relation to the place, I concede; but that one part would be with another

---

[101] St. Thomas, *Quodlibet* I, q. 10, a. 2: "Deus autem, qui est omnium causa prima, potest conservare effectus in esse sine causis proximis; unde, sicut conservat in sacramento altaris accidentia sine subiecto, ita potest conservare distinctionem materiae corporalis et dimensionum in ea absque diversitate situs."

in relation to each other, I deny. *I contradistinguish the minor.* That quantity requires parts *outside* of parts in relation to each other, I concede. But that quantity requires parts outside of parts in relation to the place, I deny. *And I deny the conclusion.* – The solution is clear from what has been said on quantity.

3rd Objection. It is impossible for two substantial forms to be in the same matter. Therefore, *a fortiori* it is impossible that two bodies be in the same place. ***Reply***. *I deny the conclusion*, for the disparity is great. The impossibility of two forms being in the same matter is derived from the very *ratio* of substantial form. For its function is to give its first *esse* and its first actuality; for this reason, all other forms that supervene only confer *esse secundum quid* and will be accidental forms. But the impossibility of two bodies being in the same place does not arise from the formal *ratio* of quantity, but only from a certain secondary effect, which God can separate from quantity.

XII. – What Impenetrability Is, Properly Speaking. Some of the more recent authors think that it is a resistive power of resistance whereby a body, through its own intrinsic action, prevents another body from invading its own place; but the Scholastics teach that it is a certain receptive potency. We adhere to this position.

For impenetrability is a property of quantity, but a resistive power is an operative quality. Therefore, impenetrability does not consist in a power of resistance, but rather in a receptive potency. Further, this resistive power is said to operate within the substance itself, among its parts. But an operation that is carried out within the substance itself and among its parts is an immanent operation. Therefore, all bodies would possess an immanent action, which nonetheless can only be ascribed to living things.

Let us conclude, then, that impenetrability is a receptive potency by reason of which two bodies are prevented from being in the same place.

The Angelic Doctor says: "That which prevents it is nothing other than the dimensions to which corporeal matter is subject, for it is necessary that that which is *per se* be the cause in the genus of each thing. But the distinction according to position belongs first and *per se* to dimensional quantity, which is defined as quantity having position; hence it is that the parts in a subject are distinct according to position due to the fact that they are subject to dimensions, and that, just as there is a distinction among the diverse parts of a body according to the diverse parts of one place through dimensions, so diverse bodies are distinguished according to different places due to their dimensions."[102]

XIII. – Whether the Same Body can be in Many Places Simultaneously. The multilocation of a body can be conceived in three ways. First, by reason of its own dimensions; second, in the mode of a spiritual and indivisible substance; third, mixed multilocation, whereby a body is in one place by reason of its own dimensions, or circumscriptively, but is present in another in the mode of a substance. All Catholic teachers unanimously acknowledge that the second and third ways are not contradictory. But the Scholastics are divided with respect to circumscriptive multilocation: St. Albert, St. Thomas, St. Bonaventure, Henry of Ghent, Capreolus, Vasquez, Ferrara, John of St. Thomas, Sylvester Maurus, de San, de Maria, Lorenzelli, and others claim that it is contradictory. Alexander of Hales, Scotus, the Conimbricenses, Suárez, Bellarmine, Valentia, Franzelin, and Pesch hold the opposite view.

XIV. – First Conclusion: "Multilocation, taken in the second and third senses, does not involve any contradictions."

Proof. For a body to be in a place in the mode of substance is nothing other than for it to exist without its own *ratio* of dimensions. But it is not contradictory for a body to be conserved without its own *ratio* of dimensions, for dimensions do not enter into the essence of a body, but are properties which God can separate from the thing. Therefore, it is not contradictory for a body to be in a place in the mode of a substance. But that which exists in the mode of a substance can be in many places; for indeed, substance includes the *ratio* whereby it can be in many places. The whole substance can be entirely in each and

[102] *Quodlibet* I, q. 10, a. 1c: "[Hoc autem] prohibens nihil est aliud quam dimensiones, quibus substat materia corporalis; necesse est enim ut id quod est per se, sit causa in unoquoque genere. Distinctio autem secundum situm primo et per se convenit quantitati dimensivae, quae definitur esse quantitas positionem habens; unde et partes in subiecto ex hoc ipso distinctionem habent secundum situm, quod sunt subiectae dimensioni; et sicut est distinctio diversarum partium unius corporis secundum diversas partes unius loci per dimensiones, ita propter dimensiones diversa corpora distinguuntur secundum diversa loca."

every place. Therefore, a body thus considered can be in many places simultaneously, as the Body of Christ truly is in all hosts simultaneously.

Thus it is not contradictory for the same body to exist somewhere according to its own appearance (*in propria specie*) and somewhere else under another appearance (*sub alia specie*), e.g., under a sacramental appearance, as the Body of Christ is simultaneously in heaven and in the Eucharist; and so it is in one place in the mode of a body, or circumscriptively, and in another place (or places) in the mode of a substance. Hence, *mixed* multilocation is not contradictory.

XV. – SECOND CONCLUSION: "It is absolutely contradictory for the same body to be in many places circumscriptively."

This is the explicit teaching of the Angelic Doctor: "To say that a body is in two places circumscriptively is to posit two contradictories together."[103]

1st Argument. It is absolutely contradictory for the dimension of one to be the dimension of two, of three, of ten, of a hundred; this would amount to the rejection of the whole of mathematics. But, if one body is in two bodies, or in three, or in ten, or in a hundred places circumscriptively, then the dimension of one would be the dimension of two, of three, of ten, of a hundred, etc. Therefore, it is absolutely contradictory for the same body to be in many places circumscriptively. *Proof of the Minor.* To be in a place circumscriptively is to apply, by way of commensuration, the dimensions of the located thing to the dimensions or surfaces of the containing body. Therefore, if one body is in two or three places, the dimension of one located body will be equivalent to the dimensions of two or three containing bodies. Therefore, the dimension of one will be the dimension of two, three, etc. *Confirmation.* That which is fully and totally exhausted in one does not retain anything that can be conferred to another. But the quantity of a located body is fully and totally exhausted in one place, such that the whole quantity corresponds to the one place and each of its parts to each of the parts of the place. Therefore, nothing whatsoever remains by reason of which the body can be in another place.

***The adversaries object*:** Indeed, the quantity of one body is exhausted naturally in one place, but God can indefinitely increase the powers of creatures, and cause one quantity to fill two places. ***Reply*:**

---

[103] Cf. *Quodlibet* III, q. 1, a. 2.

God can certainly increase the powers of creatures, so long as their essence is not destroyed. Further, if God caused one quantity to be in two places, He would destroy its essence; that is to say, He would change unity into plurality and would cause the dimension of one to be the dimension of two or of three, which is impossible.

2nd ***Argument***. If one body were able to be in many places simultaneously, it would be able to have simultaneously many positions in a place, for position follows circumscriptive position in a place. Therefore, the same body will be simultaneously standing and sitting, above and below, to the right and to the left, all of which is impossible.

XVI. – Difficulties Resolved.

1st ***Objection***. Some argue from the presence of Christ in heaven and in the Eucharist, from the apparitions of the Saints, etc. ***Reply***: None of these things prove anything. Christ is indeed in heaven circumscriptively, but in the Eucharist he is not related to the place by reason of his own dimensions; hence, there is only mixed multilocation. And apparitions are thought to be carried out by the ministry of angels, who acted the part of the Saints.

2nd ***Objection***. That which is modally distinct from something else can be separated from that thing and multiplied without that thing being multiplied. But local presence is modally distinct from located quantity. Therefore, local presence can be multiplied without quantity being multiplied. ***Reply***: It is not true that the mode can be separated from the modified thing; e.g., it does not seem possible for the roundness of wax to exist separately from a round piece of wax. But although modes in general can be removed from the thing they modify, nonetheless it does not follow that every mode in particular can be separated from the thing it modified—just as from the mere fact that an accident *as such* can be divinely held in existence without a subject, one cannot argue that every accident in particular, e.g., intellection, can exist without a subject. Further, local presence is a mode of such a *ratio* that if it were to be multiplied, a contradiction would arise, namely, if the dimension of one were also the dimension of two, of three, etc.

3rd ***Objection***. Virtual multilocation is not contradictory. But this is more difficult than circumscriptive multilocation, for it requires many miracles, while circumscriptive multilocation requires only one. *Therefore*. ***Reply***: *I deny the minor* and say against the argument given that, even though virtual multilocation perhaps requires many miracles, it nonetheless involves anything contradictory; but circumscriptive multilocation requires one miracle, but an impossible and contradictory one, namely, that unity be plurality, that the dimension of one be the dimension of two, etc.[104]

----

[104] The 12th Thomistic Thesis, approved by the Sacred Congregation of Studies, states: "The fact that a body is circumscriptively in a place, and in one place alone, no matter what power is applied to it, is brought about by the [body's] quantity" (*Eadem efficitur quantitate ut corpus circumscriptive sit in loco, et in uno tantum loco de quacumque potentia per hunc modum possit*). Cf. *Summa theologiae* IIIa, q. 75; *In IV Sent*. d. 10, a. 3.

FOURTH ARTICLE

ON SPACE AND THE VOID

I. – OPINIONS CONCERNING THE NATURE OF SPACE. The notion of space is connected with the *ratio* of place; in fact, according to many they are identical. There are almost innumerable opinions of philosophers concerning the constitution of space. Democritus, Leucippus, and Epicurus believed that place or space is the void (*vacuum*) itself, that is, a certain universal receptacle that is distinct from bodies and in which bodies move. According to some Peripatetics, space is like an immense sphere that contains all bodies. Gassendi construes it as a certain eternal, independent *ens* that is neither substantial nor accidental but that belongs to its own genus (*sui genus*) and that possesses its own dimensions distinct from the dimensions of bodies. Descartes thought that space is the same as bodies themselves, whose essence is extension. Newton and Clark seemed to confuse space with the immensity of God and time with eternity. According to Leibnitz, space is something merely relative, like time itself, for time is the order of successions, whereas space is the order among coexistents. Kant claimed that space is the mere form of external sensation. Balmes assigned extension as its *ratio*. But the Scholastics generally require two things within the *ratio* of space: extension and the relation between dimensions and that which is contained under those dimensions.

II. – FIRST CONCLUSION: "Space is not an empty void (*vacuum*), nor an independent, *sui generis*, and universal receptacle distinct from bodies."

Proof of the First Part. Space is not conceived without real extension. But an empty void, as the atomists construe it, lacks all extension. *Therefore.* Otherwise, if it were extended, the void (*vacuum*) would presuppose something else in which it itself is received, and thus there would be an infinite regress. Therefore, it is contradictory to say that space is an empty void (*vacuum*).

Proof of the Second Part. 1st Argument. If space were a universal receptacle in which extended things were contained, this receptacle, insofar as it would possess dimensions, would have to be contained in another receptacle, and this one in another, and thus there would be an infinite regress, which we already declared to be contradictory.

2nd ***Argument.*** An eternal, independent *ens* is a being that exists of itself (*ens a se*) and that is pure act. But it is absolutely contradictory for pure act to be a receptacle of bodies or a most imperfect universal potency. Therefore, Gassendi most absurdly claims that space is an eternal and independent *ens*.

3rd ***Argument.*** Whatever exists is either God or a creature. God of course enjoys the greatest actuality; and each creature is placed in a determinate category and is referred either to substance or to an accident. It is contradictory, therefore, for space to be an *ens sui generis*[105] and that is neither substantial nor accidental.

III. – SECOND CONCLUSION: "It is absolutely contradictory for space to be God's immensity."

1st ***Argument.*** Since immensity is the divine essence, space would be identical to the divine essence. But space is extended and divisible. Therefore, the divine essence would be extended, divisible, and this would undermine the divine simplicity.

2nd ***Argument.*** Space is an attribute of bodies. But an attribute of bodies cannot be said to be a divine attribute without falling into the error of pantheism. Therefore, space cannot be said to be a divine attribute, namely, the divine immensity, without the danger of pantheism.

IV. – THIRD CONCLUSION: "Space is not the order or relation among coexistents."

1st ***Argument.*** That which space presupposes is not space itself. But space presupposes the coexistence of bodies, for bodies are said to coexist precisely because they are in different points in space. Therefore, space is not the order among coexistents.

2nd ***Argument.*** If space were the relation among coexistents, those things that are attributed to space would have to be said of that relation as well. But there are many things that are said of space which cannot at all be said of those relations, for space is said to be empty (*vacuum*),

---

105 *Sui generis*, that is, belong to its own genus, or constituting its own kind. – ***The Translator.***

filled (*plenum*), equal to the located thing, etc., but it would be absurd to say that the relation of coexistents is filled, empty, etc. *Therefore.*

3rd ***Argument.*** The true notion of space requires that coexistent things be joined by a common bond. But coexistents such as Leibnitz posits cannot be joined by any common bond, for they are simple, spiritual monads that lack all extension. Therefore, the opinion of Leibnitz is by no means the true notion of space, but rather falls into the danger of idealism.

V. – FOURTH CONCLUSION: "Space is not interchangeable with bodies themselves."

The *ratio* of a container is formally opposed to that of something contained; that of something that can be occupied to that of something that occupies. But space includes the *ratio* of a container and of something that can be occupied, but body includes the *ratio* of something contained and of something that occupies, for indeed space contains bodies and is filled by bodies, whereas bodies occupy space and are contained by space. Therefore, space is formally different from bodies.

VI. – FIFTH CONCLUSION: "Space is not the mere form of external sensation."

1st Argument. Space and distance entail each other, such that a small distance and a small space are indiscriminately interchanged, as are a large distance and a large space. But it is obvious that distance, e.g., between France and America, is not something subjective or a mere effect of sensation. Therefore, space is not a mere figment of the subject.

2nd Argument. That which is purely subjective can be modified through the operation of the subject. But subjective operation is completely unable to change space; in vain does the prisoner attempt to increase, through his own subjective forms, the distance of his cell. *Therefore.*

3rd Argument. One cannot deny the reality of locomotion without denying the applied parts of mathematics and physics. But the reality of locomotion depends on the reality of space, for motion is a bringing through space. Therefore, space exhibits true and objective reality.

VII. – SIXTH CONCLUSION: "Even though space includes extension, it does not formally consist in extension."

Proof of the First Part, namely, that space includes extension. Space receives the predicates that are said of extension. For indeed it is said that great and small, narrow and wide, equal and unequal, belong to extension. But space is said to be small or great, narrow or wide, etc. Therefore, it receives the predicates that are ascribed to extension. *Further*, the proximate measure must be homogeneous with that which is measured. But the proximate measure whereby we know space is extension. Therefore, extension and space must be homogeneous; consequently, space requires extension.

Proof of the Second Part, namely that space is not formally the extension of bodies itself. Space is understood to be a certain receptacle that is *filled* and *occupied* by a body, but extension is that by reason of which the body *fills* and *occupies* space. Hence, space and extension are formally opposed as something that fills is opposed to something that can be filled, and as something that occupies to something that can be occupied. And so, space includes something beyond extension. What it includes beyond extension will be discussed next.

VIII. – SEVENTH CONCLUSION: "Space is formally constituted through the dimensions of the surrounding body insofar as in them the relation of distance is considered." In other words, in order to obtain the ratio of space, we must consider the universe in relation to its dimensions, and these dimensions are to be considered in their relation to those things which are contained under its dimensions and in relation to distance. This is the true relationship between dimensions and space.

Proof of the Conclusion. 1st Argument. We conceive space as a receptacle that contains bodies. But something is said to contain a body insofar as it surrounds and circumscribes its dimensions. Therefore, space is constituted through the dimensions of the body that surrounds and circumscribes another.

2nd Argument. The notion of a thing is obtained from its measure. But the measure of space implies the dimensions of the containing body to the contained body, for to measure space is to measure distances between the surface of the one contained body and the surface of the other containing body, or in the case of a body that is not contained by another, between the extreme parts of its surfaces. Therefore, the notion of space signifies the dimensions of the containing body in relation to the contained body insofar as in them the relation of distance is considered abstractly (*praecise*). Therefore, dimensions and distance make up space.

IX. – Place and Space. Hence we gather how space and place are related. They are not two distinct realities, for each signifies a relation to the dimensions of the containing body, but they differ in their manner of being conceived: (1) Place signifies only the first, immovable surface insofar as it circumscribes the located body; space further adds the distance among the diverse surrounding surfaces or among the parts of the surrounding surface. (2) Place signifies only length and width, but space signifies the capacity for three dimensions. (3) Space is clearly broader than place and, taken as broadly as possible, can designate the combination of all places, for place is considered with respect to a determinate body, whereas space can be seen in relation to many things—indeed, in relation to the whole universe.

X. – Division of Space. There is a distinction between (1) *pure* space and *natural* space. The first is extension itself considered abstractly, which is the foundation of geometrical ideas; hence it is also called *mathematical* space. But *natural* space, or *physical* space, is conceived as a receptacle of an extended thing or as the capacity to receive an extended thing. There is also a distinction between *absolute* space and *relative* space. *Absolute* space would be a magnitude, infinite in length, width, and depth, that [supposedly] exists before bodies and in which bodies are and move. *Relative* space is that which is circumscribed by limits.

From the above it follows that there is no absolute space. But Descartes, even though he did not believe in eternal, uncreated space, thought that the extension of the world lacked limits; hence he defends the absolute reality of space. The falsity of this opinion is immediately detected. For the world is a finite substance with respect to its essence. And infinite properties and dimensions cannot be ascribed to a thing that is finite with respect to its essence. Therefore, even if the dimensions of the world cannot be defined, they are nonetheless finite and determinate, just as the very essence of the world is bound by its limits.

(3) There is also a distinction between *real* space, *ideal* space, and *imaginary* space. *Real* space is founded on real extension; *ideal* space derives all its reality from our minds; and, finally, *imaginary* space is a

capacity to receive bodies that we can imagine as existing outside the world. Therefore, it is a certain sensible imitation of ideal space; as soon as the intellect conceives within itself ideal space, the imagination tries to express some similar representation in the sensible order. Absolute space is reduced either to imaginary or to ideal space.

XI. – The Concept of Space. As can be inferred from what was said, space has a certain objectivity, although it does not exist in the way that it is conceived. Hence, space as it is *formally* considered and with respect to the mode in which it is conceived, is a being of reason that has a foundation in reality (*ens rationis habens fundamentum in re*).

A being of reason. For space is conceived as a capacity to contain bodies that is distinct from those bodies and that has three dimensions. Further, in the natures of things there is no such capacity separately from bodies. Therefore, space is not an absolute reality.

That nonetheless has a foundation in things (*in rebus*). We observe concrete extension and three real dimensions and see that some bodies are received in others, from which we easily gather the concept of some universal receptacle of bodies, which is space considered formally.

XII. – The Notion of the Void. There are two distinct kinds of void (*vacuum*): *negative* and *privative*. The *negative* void, which is also called 'empty' (*inane*) by some, is a merely negative void in which no body is contained and which is not contained within the surfaces of some body; this is the void that they say is outside of the last heaven. A *privative* void is a place deprived of a body. They differ in many respects *a priori*: for a privative void is a place or surface that can contain but which does not contain a located body, whereas the negative void not only does not contain a body but is also not a surface or a body that can contain anything. We must also distinguish between 'void' taken broadly and 'void' understood properly. The former is that which is filled by some body that is invisible or expanded to such a degree that it no longer offers any sensible resistance, as is the air that remains under the bell of an air pump. The latter is that which lacks even the subtlest body. The present discussion concerns privative voids understood properly.

XIII. – Whether the Void Exists in Nature. The materialists—Leucippus, Democritus, and certain modern physicists—

defend the existence of the void. It is certain, nonetheless, that the void does not in fact exist in nature.

1st Argument. Light, attraction, and the other corporeal powers are propagated throughout all the parts of the universe. But the propagation of a corporeal power cannot occur through the void. Therefore, there is no void. *Proof of the Minor*. The propagation of a corporeal power, as an accident, requires a corporeal subject. But an absolute void is the exclusion of a corporeal subject. Therefore, it would be a contradiction for the propagation of a corporal power to occur through the void.

2nd Argument. It is evident that light is diffused in a straight line. But, if there were a void within the pores of a body, the diffusion of light would not occur in a straight line; for since corporal light could not roam outside of a corporal subject, it would always have to follow some corporeal subject, and it would thus go across innumerable circuits. Therefore, there is no perfect void within the pores of a body.

– ***Objection***: How else, then, can the barometric void (*vacuum*) and the void of an air pump or the void that is said to be above our air be explained? ***Reply***: It is not a true and perfect void, but a relative void, a 'void' improperly speaking. When air can be expanded to such a degree, it no longer retains a sensible resistance; but yet, under the bell of an air pump and within the barometer, some subtle body remains. For indeed, light is propagated through it. And light requires a corporeal subject. Therefore, there remains a corporeal subject; nor can one infer the existence of a perfect void. Further, above our air there is a very tenuous body by reason of which the powers of the celestial bodies are propagated.

XIV. – What is to be Said of the Dictum: "Nature abhors the void." – Although many have abused this principle to explain the rise of water in pipes, nonetheless it can be accepted in a legitimate sense. Since nature seeks its own good, it shuns those things that are contrary to its own good. But the void prevents the good of nature, that is, the continuity and the conjunction of bodies, which is necessary to nature for bodies to interact, for light and the other powers to be able to be diffused, and especially through the shortest distance, which is in a straight line. Therefore, it is very true that nature abhors the void.

XV. – Whether the Void Can Exist at Least by God's Intervention. Therefore, since the void is contrary to nature, it must be concluded that the void neither exists naturally nor is naturally possible. The Scholastics most commonly (*communius*) teach also that the angel cannot cause the void because the inclination of nature to resist the void is stronger than its inclination to obey the angel. But absolutely speaking, can God not induce the void? Descartes and many Scholastics, e.g., Albert Farges,[106] denies this; but many Scholastics affirm this, both among the ancients, such as John of St. Thomas and Goudin, and among the more recent, such as De San and De Maria.

– Goudin argues thus: "God can at least destroy the whole universe and conserve a single man; who will dare to deny this power to God? But this being the case, there will manifestly be a distance between the man's two feet and between his two hands, even though there is no body in between. Therefore, there can be a distance between the extremes that is not filled by any intermediary body."[107]

An absolute void, a *spacious nothing* such as the ancients posited, is without a doubt contradictory; but to us it does not seem contradictory to posit a privative vacuum, namely, a place that is not filled by any body. For indeed, place and space are not intrinsically located, but rather, those terms denominate extrinsically. But, if they are located extrinsically, absolutely speaking place and space can be located things without ceasing. Therefore, absolutely speaking there can be a place that is not filled by any body or a privative void.

> "The void is nothing other than a place without space, insofar as it is able to contain within itself some body but does not contain actually any body within itself. But it is manifest that it is not intrinsically contradictory for there not to be any body between two surfaces that are actually distant from each other."[108]

----

[106] *L'Idée de continu*, 3e partie, n. 3.

[107] *Physic*. q. 4, a. 3: "Saltem posset Deus destruere totum universum et unicum hominem conservare; quis hoc Dei potentiae audeat denegare? Hoc vero dato casu, manifeste dabitur distantia inter ambos pedes et ambas manus, quamvis nullum inter illa mediet corpus. Ergo potest dari distantia inter extrema nullo medio corpore repleta."

[108] DE SAN, *Cosmol*. n. 454: "Nihil aliud est vacuum quam locus sive spatium, quatenus est quidem capax continendi intra se corpus aliquod, actu tamen nullum corpus intra se continet. Non autem intrinsice repugnare ut inter duas superficies quae actu inter se distent nullum corpus actu intercipiatur, manifestissimum est."

FIFTH ARTICLE

On duration in general[109]

I. – *Ratio* of the Article. Space and time are almost conceived together in the mind. Each depends on motion. For space is required for a body move from one terminus to another; but for it to move successively, it requires time. The consideration of time, therefore, logically follows the consideration of space. But since body is a species of duration, certain things must be examined first concerning duration.

II. – The Notion of Duration. Duration is unanimously (*communiter*) defined as "the permanence of a thing in *esse*." For as long as a thing is said to last (*durare*), it is said to be in act. An instance, therefore, does not properly have duration because it does not persevere but is the terminus of duration, just as the point is not extension, but the terminus of extension. But can a lasting thing (*res durans*) be said to have duration in the first instance in which it exists? – It does not have complete duration, properly speaking, because it does not yet attain perseverance or continuity in *esse*; but nonetheless, because the perseverance in being is already beginning, in its first instant the thing can be said to last inchoately.

Now, a thing can remain in being in three ways: (1) There is a type of permanence whereby a thing is altogether immutable in its being, both with respect to its substance and with respect to its operation; this permanence in *esse* is called *eternity*. (2) There is a type of permanence whereby a thing is immutable with respect to its substance but not with respect to its operation, and this permanence is called *aeviternity* (*aevum*). (3) There is a type of permanence in *esse* whereby a thing is mutable both with respect to its operation and with respect to its substance, and this duration is called *time*.

---

[109] On duration and time, one can refer to the following authors: St. Thomas, *Summa theologiae* Ia, q. 10, and his commentators: Cajetan, Bañez, Buonpensiere, etc. and generally the Scholastic philosophers already cited, in their *Physics*. See also Augustine, *Confessiones* 11 and 14; Boethius, *De consolatione philosophiae* 5; Balmes, *Philos. fondament.* 7; Farges, *L'Idée de continu*; Nys, *La notion de temps*.

III. – WHETHER DURATION IS DISTINCT FROM THE LASTING THING (*RES ENDURANS*). It is certain that in God there is no real distinction. And so, in God, the divine essence is not distinct from its duration, but rather, God is His own eternity. But the Scholastics dispute whether duration in creatures is different from their essence. Scotus, Suárez, and Vasquez deny that it is. But many others defend a real distinction between duration and essence; in fact, many claim that duration is distinct from the existence of the lasting thing (*res endurans*).

IV. – CONCLUSION: "In creatures, duration is really distinct from the essence of the lasting thing; in fact, it adds something beyond their existence, but something merely extrinsic."

Proof of the First Part. Duration is permanence in *esse*. But it befalls (*accidit*) the creature to have *esse* and to remain in *esse*, for it can be reduced to nothingness. Therefore, duration does not belong to the essence of the creature, just as its existence is not interchangeable with its quiddity.

Proof of the Second Part, namely, that duration adds something extrinsic beyond existence. It is manifest that it does not add any intrinsic entity. For if duration added something intrinsic beyond *esse*, then it would no longer be a mere continuation in being, but a new production. – Nonetheless, we now briefly show that it adds something extrinsic. Duration is a continuation in being. And a continuation in being connotes an action that continues and gives *esse*. Therefore, duration adds the extrinsic connotation of an action that conserves *esse*. Hence, the more the conservation depends on defectible causes, the more defectible is the duration.

In successive things, which are said to last (*durare*) by reason of a continuous flow, insofar as one part ceases to be and another begins to be, duration occurs through the addition of existence to existence; and thus the whole successive existence is not distinct from the whole duration, but one part is distinct from another. This is the view of John of St. Thomas.[110]

V. – THE TERMINUS OF DURATION. In the duration of an imperfect thing, which is subject to beginning to exist and to ceasing to exist, there is a distinction between two termini, namely, a *beginning* and

[110] *Philos. Nat.* I. P., q. 18, a. 1.

end. Further, these termini are altogether indivisible. For whatever is divided itself has a beginning and an end; for this reason, in order to avoid an infinite regress, something that functions as a terminus must be altogether indivisible. Hence, just as in extension there are indivisible points that begin and end a line, so in duration there are indivisible instants that begin and end duration.

Philosophers, however, distinguish between *intrinsic* and *extrinsic* instants. An *intrinsic* instant is that within which the thing is present, namely, that which serves as the beginning or ending terminus in such a way that the thing is present in it according to its essence. But an *extrinsic* instant is that within which the thing cannot be present according to its essence; namely, that which serves as the beginning or ending terminus in such a way that the thing is not present in it, but is present only in the whole duration that follows—if it be the terminus that begins the thing—or was present in the whole duration that preceded—if it be the terminus that ends the thing. Further, the intrinsic end can be considered in two ways: the instant of beginning to be at which the thing is first present, since it was not present in the whole duration that preceded the instant, is called *the first being of the thing* (*primum esse rei*); whereas the instant of ceasing to be at which the thing is present for the last time, such that it will not remain in the whole duration that follows the instant, is called *the last being of the thing* (*ultimum esse rei*). The extrinsic instant is also twofold: the instant of beginning to be at which the thing is not yet—but will be present in the whole duration that will immediately follows the instant—is called the *last non-being of the thing* (*ultimum non esse rei*); whereas the instant of ceasing to be at which the thing is no longer present—but was present in the whole duration that immediately preceded the instant—is called *the first non-being of the thing* (*primum non esse rei*).

VI. – WHICH BEINGS BEGIN TO BE AND CEASE TO BE THROUGH AN INTRINSIC INSTANT. (1) Successive beings begin to be and cease to be through an intrinsic instant. For those things are considered to begin to be through an extrinsic instant whose essence is not present at the instant of beginning to be, and those things are said to cease to be through an extrinsic instant which no longer exists in the instant of ceasing to be, but were present in the whole duration that preceded the instant. But the essence of successive things cannot be present at the instant of its beginning to be or at the instant of its ceasing to be, for succession, which implies a before and an after, is not present in an instant, which does not have before or after. Therefore, successive beings, like motion and time, begin to be and cease to be through an extrinsic instant.

(2) Permanent beings, like substances, begin through an intrinsic instant. For indeed, those things are said to begin to be through an intrinsic instant whose whole essence is present at the first instant. But since the essence of permanent things is altogether indivisible, it can be present in the first instant of its production. Therefore, permanent things begin to be through an intrinsic instant.

(3) Permanent beings *in fact* (*de facto*) cease to be through an extrinsic instant; but absolutely speaking, it is not a contradiction for some permanent being to cease to be through an intrinsic instant.

Proof of the First Part of the Assertion. The ceasing to be or corruption of one substance occurs for the sake of the beginning to be or generation of another; hence the same instant at which the duration of the prior substance ends is the instant at which the subsequent substance begins, for we ascertain that substances begin in an intrinsic instant. Therefore, it cannot be intrinsic to the thing that ceases to be. For the form that ceases to be is not present at that instant; otherwise there would be two substances in the same matter at the same instant. Therefore, substances cease to be through an extrinsic instant.

Proof of the Second Part. The reason why a substance ceases to be through an extrinsic instant is that it does not cease to be *per se*, but for the sake of the generation of another. But absolutely speaking, God can cause a substance to cease to be *per se,* e.g., that an angel be annihilated. Therefore, some substance could cease to be through an intrinsic instant; for, since it is indivisible, its whole essence could be present in the last instant and altogether cease to be instantly afterwards.

VII. – THREE KINDS OF DURATION. We already distinguished between three species of duration: *eternity*, *aeviternity* (*aevum*), and *time*. But some claim that the difference between these is as follows: eternity lacks both a beginning and an end; aeviternity has a beginning but no end; and time has both beginning and end. This difference is *per accidens* and not *per se*. For if time always had been and would always be there would still remain a difference between eternity and time, for eternity would be the measure of a permanent being whereas time is the measure of motion. This opinion, therefore, in no way explains the essential difference between eternity and time. But it is true in one respect. For indeed, in duration two things are to be distinguished: the measure and the measured thing. If this opinion be considered with respect to measure, it will exhibit something true because that alone which has a beginning and end in time is measured by time.

– Similarly, this opinion does not assign an essential difference between eternity and aeviternity. For there it would not be a contradiction for permanent and aeviternal things always to have been, as we have shown in the first treatise of *Natural Philosophy*;[111] also, an aeviternal thing could cease to be, e.g., if God annihilated it. Therefore, there could be an aeviternity that lacked a beginning but which had an end.

And so, the essential difference is this: Eternity is the duration of something that is in all respects immutable both with respect to its substance and with respect to its accidents; aeviternity is the duration of a thing that is immutable with respect to its entity but mutable with respect to its accidents; and, finally, time is the duration of a thing that is in all respects mutable both with respect to its substance and with respect to its accidents.

VIII. – ETERNITY. Boethius defines it as "the whole, simultaneous, and perfect possession of interminable life."[112] It is called a '*possession*' because in eternity nothing is expected as future, but all is present in a permanent and restful possession. – '*Whole*' is added, not because it has parts, but insofar as nothing is lacking in it. It is '*simultaneous*' and '*perfect*' because in time, as we shall soon declare, two things are to be considered: namely, *time* itself, which is successive, and the *now* of time, which is imperfect. Hence it is said: '*whole and simultaneous*' in order to rule out time, and '*perfect*' to exclude the now of time.

Therefore, since eternity is opposed to time, excludes all succession. For indeed, true succession implies change, in which there is a before and an after. But eternity is the duration of an *ens* that is altogether immutable. Therefore, it excludes all succession.

– It is called '*life*' because an eternal subject is not only an *ens*, but a living thing. Further, life extends itself somehow to operation but not

---

[111] Q. 3, a. 2.

[112] *De consolatione philosophiae* 5.6 (*PL* 63, 858): "Interminabilis vitae tota simul et perfecta possessio." – ***The Translator.***

to *esse*. But the extension of duration seems to have to do more with operation than with *esse*.

It was better therefore to define it as "*the possession of life*" than as "the possession of being." – Finally, '*interminable*' is added because just as a being that is altogether immutable does not have succession, so it does not have a beginning or an end.

– Theologians distinguish between *participated eternity* and the *mode of eternity*. Participated eternity is a certain measure of created operation, but which does not admit of variation, as are the beatific vision and beatific love. But the mode of eternity is the measure of a supernatural thing, as are the infused virtues, which, having been derived immediately from God, participate somewhat in His duration.

IX. – Aeviternity. We defined aeviternity (*aevum*) as the duration of a thing that is immutable with respect to its entity. Because substance properly has entity, aeviternity is properly the measure of an incorruptible substance. But yet, accidents that are altogether incorruptible, such as the intellect and the will, are measured by aeviternity.

The mode of aeviternity is the measure of some natural operation which is nonetheless indefectible; for example, an angel always has knowledge of its own essence; this operation is a mode of aeviternity.

It is disputed whether in aeviternity there is succession. St. Thomas, Suárez, and many others deny that there is; St. Bonaventure, the Conimbricenses, and Sylvanus Maurus affirm it.

X. – Conclusion: "In aeviternity there is no intrinsic or real succession; however, a certain virtual and extrinsic succession can be admitted."

Proof of the First Part. There cannot be true succession, that is, a true distinction between before and after, unless that which is before ceases to be and that which is after begins to be. But that variation of beginning to be and ceasing to be cannot be conceived in an aeviternal substance. Therefore, in an aeviternal substance there is no true succession. *Proof of the Minor*. Since an aeviternal substance cannot be transformed into another, its variation would consist in the fact that in each moment it ceased to be and was produced again; but in this case the divine operation in virtue of which the substance lasts (*durat*) would not be conservation but repeated creation. Therefore, the succession of before and after would not befall to the same substance, but to diverse substances.

Proof of the Second Part. An aeviternal substance does not have possession of all of its existence, for it depends on God in such a way that it may not have been created and such that it could even now be reduced to nothingness. Hence, if God had not made a certain angel from the beginning, but had made him today, then that angel would have a shorter duration (*minus duraret*) than one created from the beginning; and if an angel were to be annihilated, it would have a shorter duration than one who will always exist. Therefore, the duration of an aeviternal substance can be said to be virtually and extrinsically successive by reason of its dependence on a free, conserving cause.

An aeviternal substance also has succession extrinsically "insofar as it does not coexist with all the parts of time simultaneously, but only according to their changes. Even though in itself it is something fixed and not transient, nonetheless, given its limitation, it does not coexist with all of them but awaits things that come and go; just as the tree that is fixed in a river does not coexist simultaneously with all the parts of the flowing water, but coexists with them according to succession, not its own succession, for the tree is fixed, but the succession of the water which comes to it, and thus awaits the parts of the water so that it can coexist with them, one at a time. But eternity proper lacks this extrinsic succession due to its infinity and immutability not only in *esse* but also in measuring, and does not await for things with succession to flow and thus to coexit with it, but rather it contains them immutably before they exist mutably in themselves, through a superior mode of measure from which they are derived—as if there were a tree of such magnitude that it occupied all of the water of the river and would coexist with all its parts, even if they had succession in themselves."[113]

---

[113] JOHN OF ST. THOMAS, *loc. cit*, art. 2: "[Q]uatenus non simul coexistit omnibus partibus temporis succedentis, sed secundum earum mutationem. Licet in se sit aliquid fixum et non transiens, tamen propter limitationem suam non omnibus simul coexistit, sed expectat res quae succedunt; sicut arbor fixa in fluvio non coexistit simul omnibus partibus aquae fluentis, sed coexistit illis secundum successionem, non suam, quae fixa est, sed aquae, quae illi advenit, et sic expectat illas ut illis possit coexistere. At vero aeternitas propter suam infinitatem et immutabilitatem non solum in essendo, sed etiam in mensurando, caret ista succesione extrinseca, et non expectat res successivas ut fluant, et tunc illi coexistant, sed immutabiliter eas continet, antequam mutabiliter in se existant, per

XI. – Whether in Aeviternity there is a *Ratio* of 'Becoming Old' (*veteratio*) and 'Becoming New' (*innovatio*). Certain theologians have thought that there is succession without 'becoming new' and 'becoming old', and vice versa. But it should be acknowledged that 'becoming new' and 'becoming old' are the same as succession of the prior and the posterior. For, since the prior and posterior cannot be simultaneously, it is necessary that the posterior come as something new and that later it recede and cease to be. But to come as something new is to 'become new' (*innovari*), and to cease to be and to recede is to 'become old' (*inveterari*). Therefore, there is no succession of prior and posterior without 'becoming new' and 'becoming old', nor can 'becoming new' and 'becoming old' be conceived without the succession of the prior and the posterior.

Therefore, since aeviternity excludes succession, it also lacks all 'becoming new' and 'becoming old'.

We leave all other issues concerning eternity and aeviternity to theologians.

---

modum superioris mensurae a qua derivantur. Sicut si esset arbor tantae magnitudinis, quae occuparet totam aquam fluvii, simul omnibus partibus coexisteret, licet illae succederent inter se."

SIXTH ARTICLE

On time

I. – Definition of Time. Time was already defined as "the duration of a mutable being with respect to its accidents and its substance." But there is a better known definition given to us by Aristotle: "The numbering of motion according to before and after."[114] It is called a *numbering*, that is, a measure, for number is the measure that is better and more frequently known to us. Further, it is not the measure of just anything, but of motion; Aristotle understood it of the motion of the first mobile being (*primum mobile*), but it can also be taken as the motion of any mobile being, properly of locomotion and, in a sense, of spiritual motion.

Aristotle says *according to before and after* in the sense that in motion there are parts that flow successively, of which one is always prior in duration and delay and the other is posterior; and the prior part is called *past* (*praeterita*, lit. "gone beyond") and the posterior is called *future* (*futura*, lit. "about to be"), which parts we number, and for the measure of all durations we determine how long or brief they are.

II. – The Aforesaid Definition is Further Explained.[115] We should first consider the fact that time is not motion; for indeed motion is fast or slow, but time is not fast or slow *in itself*, even if someone should think it is *subjectively* slow or fast. Many motions are measured by time, but nothing is its own measure. Therefore, time is not motion. For the same reason it is evident that time is not a succession of motions, for succession can be faster or slower, which means it is measured by time. Time is also not something mobile because the parts of something mobile are in act simultaneously, whereas the parts of time exist in act only successively.

– Although time is not motion, nonetheless it is not without motion. For indeed, that is not without motion which cannot be conceived without motion. But time cannot be conceived without motion. Therefore, time is not without motion. The minor is evident

[114] *Physics* 4.11.

[115] Cf. St. Thomas, *In IV Physic.*, lect. 17.

through experience. Those who vigorously apply the whole attention of their mind to something, or those who sleep, or those who exist in a catalepsy do not apprehend time, because they do not experience time, that is, an intermediary successions, but join the end of one thing with the beginning of the following thing, skipping intermediary succession. They only apprehend time, then, when they perceive motion and in it distinguish before from after. Time, therefore, is not without motion.

It remains for us to inquire whether time follows motion by reason of motion absolutely or by reason of the before and the after. But it is clear that it is by reason of the before and the after because we perceive time due to the fact that we apprehend succession in motion, and in it we distinguish the before from the after.

Now, there are three main kinds of motion, namely: motion of *alteration*, motion of *augmentation*, and *locomotion*. But time, insofar as it is continuous and uniform, does not belong to motion of alteration or of augmentation because these motions are not in all respects continuous and uniform. It remains, therefore, that time belongs to locomotion, namely, that it is its numbering. For through number we judge something as more or less, and by time we judge motion to be greater or less. Time, therefore, is a number. And further, number is twofold, namely: *absolute* number, the kind whereby we enumerate things, such as one, two, three; and *numbered* number, such as ten men. Time, of course, is not a numbering number, but a numbered number, for it is a before and an after, as numbered things are in motion.

Time participates both in permanent quantity and in discrete quantity according to different respects. Insofar as time is something continuous whose parts are joined by a common terminus, one can ascribe to it some of the predicates of permanent quantity; for this reason it is considered to be long or brief, just as a line is said to be long or brief. Moreover, time is conceived as number, and thus it has the properties and predicates of number, in which we can speak of little time or much time. *Simpliciter*, however, it is not discrete number because its numerable parts are not distinct in act, but it is a *flowing* quantity.

III. – The Uniformity of Time. Time has continuity and uniformity which is based on the continuity and uniformity of motion.

Therefore, wherever there is uniform motion there can be time; hence if the lights of the heavens ceased to shine and the potter's wheel were to be moved, there would still be time, as Augustine explains. But yet, because the motion of the stars is more stable and less irregular, the duration of the motion of the stars, namely, twenty four hours, is taken as a uniform measure, especially the duration of the Earth's revolution about the Sun. But recall that the motion of the heavenly bodies is not completely uniform, for the planets move more quickly as they approach the Sun. Therefore, we must further seek some first mobile being (*primum mobile*) that serves as the center of the whole system of the universe and whose completely uniform motion is the rule and measure of all others. By comparison to this *primum mobile* we could obtain the absolute unity and uniformity of time, just as the absolute immobility of place can also be established in relation to some body that is completely immobile and which contains all places and located bodies. But since it cannot be established that such *primum mobile* exists, we use an *abstract* and *mathematical* time, which scientists use as the absolute measure of all motions.

IV. – Division of Time. There is a distinction between *intrinsic* and *extrinsic* time. The former is the intrinsic duration that is proper to each motion; the latter is the duration that is used to measure other durations and which, thus, is common to many things.

Now, extrinsic time is subdivided into *general* and *particular*. *General* time is the celestial motion that is used to measure all motions; *particular* time is a duration that is used to measure some motions, such as the motion of a clock.

One could also distinguish between *imaginary* time and *real* time: *imaginary* time is the number of a possible motion according to before and after; *real* time is the numbering of real motion.

Cosmographically, time is divided into *astronomical*, *true*, *mean*, and *sidereal*. *Astronomical* time is that whereby we count twenty four hours without interruption, beginning from the meridian, whereas *true* time is that which is determined by the return of the Sun to the meridian. But because the motion of the Sun in a straight line varies, true time is not always fixed. Nonetheless, it was necessary to establish a fixed measure for clocks, and for this reason there is said to be *mean* time which is always divided into equal parts, namely, into twenty four equal hours. True time, therefore, is sometimes longer and sometimes shorter than *mean* time. And finally, *sidereal* time is that which measures the motion of some star or its return to the meridian. It is a bit shorter than mean time.[116]

V. – What Kinds of Things are in Time. One thing is to exist '*with time*', or 'during time', and another thing is to exist *in time*. For God and the angels are 'during time', or while time exists, but they are not in time. For something to be properly said to be subject to time, it must be in it either as something included is in that which includes it, as days are in a year, or as a measured thing is in a measure. All those things whose entity is mutable, and consequently, measurable by time, are subject to time. Hence, any sensible motion is subject to time; and the mobile thing itself, or the mobile substance, is measured by time.

But the Scholastics dispute whether a corruptible substance is measurable immediately in itself and not only by reason of motion. Scotus, the Conimbricenses, and Suárez hold that corruptible substances are not measured by time.

They offer the following argument: Time is a successive duration; but substances are not successive, but permanent. Therefore, they are not measured by time.

Others, however, as John of St. Thomas, resolve the question thus: the substance of a corruptible thing is not measurable by time according to the concept of substance but only insofar as it has quantity of duration or insofar as it is permanent in mutable rest. We adhere to this opinion.

Proof of the First Part. A mobile thing does not have before or after with respect to the concept of substance, but is abstracted from motion and succession altogether. Therefore, it can only be measured by time insofar as it has the quantity of duration; for measuring is due to quantity. Therefore, that is properly measured by time whose quantity is measured by time.

---

[116] Bear in mind that the amount of daylight varies from Winter to Summer, and that before the invention of the mechanical clock, daylight was always divided into twelve hours, irrespective of the season. Thus, the shorter day of the Winter was divided into twelve equal parts, and the longer day of the Summer was also divided into twelve equal parts. These twelfths of a day are called hours, irrespective of the season. Therefore, the Summer hour was much longer than the Winter hour. These hours, which last a varying length of time, represent real time. It is the hour of average length—the hours of the Spring and Fall equinoxes—that are called mean time, and it is on the basis of mean time that modern clocks keep time. – ***The Translator.***

Proof of the Second Part, namely, that corruptible substance is measured by time not as far as its substance but as lasting and permanent in a mutable rest.

The measure of a permanent thing and that of a thing at rest are the same. But the rest of a mutable thing is measured by time. Therefore, the permanence of a mutable thing, or the substance itself as permanent, is measured by time. *The major stands*, for 'permanent' and 'at rest' are the same, since to be at rest means to remain (*permanere*) as before. *Proof of the Major*. The measure of opposites is the same. But rest is the opposite of motion. Therefore, time, which measures motion, is also the measure of rest—it is *per se* the measure of time, whereas it is *per accidens* the measure of rest.

John of St. Thomas responds to the argument of the adversaries: A permanent thing, even if it is not successive formally and in act, is nonetheless radically and extrinsically successive due to the fact that, from its root, it seeks being able to be moved and corrupted, and depends on mutable causes.[117]

Therefore, those things that are composed of matter and form, even man himself, are subject to time. The human soul, reduplicatively[118] as the form of the body, can be said to be in time because it is the part of a mutable composite; but as an immaterial and incorruptible substance it is not measured by time but by aeviternity.

The sense operations of the human soul are in time; intellectual operations, objectively considered in its relation to phantasms, are also subject to time; for indeed, they depend on the senses and on the

---

[117] *De tempore*, a. 3.

[118] *Reduplicatively*, i.e., *qua* itself, or in the respect in which it is named. For example, "man reduplicatively" means "man inasmuch as he is a man" or "man *qua* man"; this is to be distinguished from non-reduplicative uses of "man", for example, "man insofar as he is an animal." There are things that are true of man reduplicatively that are not true of him non-reduplicatively. For instance, that "man reduplicatively as man is rational" is true, for being rational is part of what it means to be a man, but that "man insofar as he is an animal (non-reduplicatively) is rational" is false, for being rational is not part of what it is to be an animal. This term has much theoretical mileage for the Thomistic tradition. For example, whereas the object of all sciences is *ens*, albeit in different respects, only metaphysics has as its object *ens* taken reduplicatively, that is, *ens qua ens*, or *ens inquantum hujusmodi*. (St. Thomas does not use the term 'reduplicatively', but uses the circumlocution *inquantum hujusmodi*, 'insofar as [it is] of its kind' or simply '*qua* itself'.) – ***The Translator.***

imagination, which are measured by time. Hence, it is commonly said that we require time to think because time is necessary for us to form the phantasms that we use for thinking.

But intellection considered subjectively, insofar as it is altogether spiritual, as are also the operations of the separated soul or of the angels, are not properly subject to our time. For this reason, the Scholastics admit that there is in souls and in angels some sort of time that has a different *ratio* from our time. For time is the measure of motion. But the motion of spiritual things has a different *ratio* from the motion of corporeal things. Therefore, in spiritual creatures there is a time which is different from our time.

Now, the time that measures spiritual operations is not continuous, but discrete. For indeed, continuous things are those which are joined by a common terminus. But spiritual operations that occur in succession are not joined by a common terminus, like the parts of quantity, but are altogether distinct and discrete. Therefore, there is no continuous time in thoughts and spiritual operations. Nonetheless, if an angel moves in a continual motion, as when they continuously move or transfer some body, then angelic time can be said to be continuous. This is the unanimous opinion of the Scholastics, especially of the Thomists.

VI. – The Elements of Time. Time, properly speaking, is a species of continuum. We already have shown in article II, however, that the continuum is made up not only of indivisible parts, but also of divisible parts that are joined through indivisible parts. In time, therefore, there are both divisible and indivisible parts: the divisible parts are called *before* and *after*, whereas indivisible parts are called *instants*. And thus, it is clear that the before and after of time are joined by indivisible instants. And, indeed, time is a successive and yet continuous *ens* which implies a continuous flow from before to after. But it would not be a continuous flow, but would be interrupted, unless the before and after were joined by some common bond. Therefore, in time there is some bond that joins the before and the after. But the bond that joins the parts of the continuum must be indivisible, otherwise the parts of the bond itself would be joined by some other bond, and thus we have an infinite regress. Therefore, the before and after of time are united by something simple and indivisible, which is the instant. The before is called the *past* (*praeteritum*, lit. "gone beyond") because it already was, and the after is called the *future* (*futurus*, lit. "about to be") because it is not yet; and the copulative instant is called the *fleeting present* (*nunc fluens*, lit. "flowing now") because it perpetually varies in contrast to the *eternal present* (*nunc stans*, lit. "standing now"), which is invariable and constitutes eternity. Therefore, *past*, *future*, and *fleeting present* are necessary elements of time.

VII. – WHETHER TIME IS PRESENT BY REASON OF A [DIVISIBLE] PART OR BY REASON OF AN INSTANT. This question is very difficult to resolve. On the one hand, it is unintelligible how a successive being exists by reason of parts that do not exist; on the other hand, it is not evident how parts can be past unless they existed at some point in time, or even how an instant makes the future exist, since it presupposes that the future does not yet exist, i.e., is not yet present, as John of St. Thomas rightly notes. The philosophers, therefore, are divided. Durandus, the Conimbricenses, and Suárez hold that time is present, not only by reason of the instant but also by reason of a [divisible] part. The Thomists and the greater part of the Scholastics hold the opposite view.

VIII. – CONCLUSION: "Time is not present and existing by reason of a [divisible] part insofar as it is a [divisible] part, but only by reason of an indivisible instant."

1st Argument. Time is a successive *ens*. But it is contradictory for a successive being to exist and be present according to some determinate, divisible part, no matter how brief. Therefore, it is impossible for time to exist and to be present according to some determinate, [divisible] part. *Proof of the Minor.* It is absolutely contradictory for many successive parts to exist simultaneously, in fact simultaneity and succession destroy each other. But, if a successive being existed according to some divisible part, then many successive parts would exist simultaneously. *Therefore. Proof of the minor of the subsumed argument* that is, of the premise that if a successive being existed according to some divisible part, then many successive parts would exist simultaneously. There cannot be something present by reason of itself unless all the parts of which it consists exist simultaneously. But any part of a continuum is composed of other parts into which the part is itself divisible; for the continuum is always divisible into parts that are themselves divisible. Therefore, there cannot exist a continuum unless many parts exist simultaneously, either permanent parts in a permanent continuum or successive parts in a successive continuum. Therefore, if a successive being could exist according to some [divisible] part, then many successive parts would exist simultaneously.

2nd ***Argument.*** There is nothing present in time except what is *now*; for what existed *before* is not present, but went away; and what will exist *after* is not present, but is to come. But *now* is not a [divisible] part of time, since it cannot be divided into before and after, but it is an indivisible [part] of time. Therefore, time is not present by reason of a [divisible] part, but by reason of something indivisible.

– ***We reply to the argument of the adversaries***: The parts of the past existed and were present, not by reason of themselves but by reason of an indivisible instant; and the parts of the future become present initially by reason of the initiative from which they begin, and are made to exist consummatively by another terminative instant.

IX. – Other Difficulties Resolved.

1st ***Objection.*** The parts of time are divisible, but an instant is indivisible. But divisible parts cannot exist in an indivisible [part]. Therefore, the parts of time cannot exist and be present by reason of an instant. ***Reply.*** *I distinguish the minor*: that indivisible parts cannot exist in an indivisible part as contained therein, I concede; but that indivisible parts cannot exist in an indivisible part as joined to it, I deny. And in the same way, *I distinguish the conclusion.* – The instant, therefore, makes the [divisible] parts of time exist and be present insofar as, by finishing the past part and beginning the following part, it joins them and thus makes time exist, just as the union that joins a link with another makes the chain exist, as Goudin says.[119]

2nd ***Objection.*** If parts exist by reason of an instant, before and after are joined by the same *now*. But those things that are joined by the same *now* are simultaneous. Therefore, before and after will exist simultaneously, which is absurd. ***Goudin replies***: Even though the same *now* joins before and after, nonetheless, it affects them differently. For it ends the before and it begins the after; and the fact that the same 'now' ends one thing and begins the next does not mean that the two things exist simultaneously, but for the one succeed the other.[120]

---

[119] *Phys.* De tempore, a. 2.
[120] *Ibid.*

X. – The Concept of Time: Whether Time is Something Real or Only Something that Belongs to Reason. From the foregoing we can conclude that time considered *formally* is a being of reason with a foundation in reality (*ens rationis cum fundamento in re*); in other words, time is real with respect to the thing conceived, but is a formality of reason with respect to its mode of being conceived.

Proof of the First Part. Time is a successive *ens*, or the measure of motion according to before and after. And a successive being exists in reality, for motion lasts successively in reality, and experience attests that progress in a journey is not accomplished suddenly, but gradually and successively. Therefore, time with respect to the thing conceived, or with respect to its foundation, exists in reality.

Proof of the Second Part. Time with respect to its formal [*ratio*] is conceived as having the *ratio* of a measure. But it can only have the *ratio* of measure through an act of the intellect. Therefore, time considered with respect to its formal [*ratio*] is found only in the intellect. *Proof of the Minor.* A measure implies two things: simultaneity, or the collection of parts into one, and the application to that which is measured. But the collection and simultaneity of the parts of time is not in the nature of things but is done through the soul, which gathers the parts and compares the past with the future; and thus, the application of duration to that which is measured is done by the soul. Therefore, time has the *ratio* of measure only through an act of the intellect.

– Each part is confirmed by the authority of St. Thomas: "Time has a *foundation* in motion, namely, the before and after of motion itself; but with respect to that which is *formal* in time, namely, numbering, it is completed through the operation of the intellect that enumerates."[121] And in the reply to the second argument of that article, he refers to and approves the Philosopher's statement that "If there were no soul, there would be no time."[122]

---

[121] *In I Sent.*, d. 19, q. 5, a. 1: "[Tempus] habet fundamentum in motu, scilicet prius et posterius ipsius motus; sed quantum ad id quod est formale in tempore, scilicet numeratio, completur per operationem intellectus numerantis." Either erroneously or on the basis of a different edition, Hugon cites q. 10, a. 4, which in our edition (Mandonnet) does not exist – ***The Translator.***

[122] *Ibid.* q. 2, a. 1c: "Si non esset anima, non esset tempus." Note that in the Mandonnet edition this quote does not come from the responses of the article cited above, but from the *corpus* of an earlier question. – ***The Translator.***

XI. – Whether Time is Something Absolute or Relative. It is the opinion of some that time is only relative. For, indeed, time is measured by clocks. But a clock is an arbitrary and relative measure in which there can be acceleration or retardation of time; in fact, even the motion of the Sun itself, according to which clocks are regulated, can be accelerated or retarded. Therefore, they say, time is relative.

Nevertheless, we should acknowledge that time exhibits something absolute. For even though the measure we use is arbitrary, duration itself is absolute and immobile, as it is evident that an hour is a certain absolute quantity, which is its own nature. If the motion of all clocks or of all the heavenly bodies were retarded, it would still be true that the duration of one hour remains the same, that a quarter hour is fifteen times greater than a minute and that an hour is four times greater than a quarter hour.

Therefore, just as space is an absolute quantity, even if we measure it with an arbitrary measure, so time is in itself something absolute, even if the measure whereby we reckon it is arbitrary.

XII. – Refutation of False Opinions. Having fully understood that which we have hitherto discussed, we can now easily refute other opinions on the nature of time. The opinion of Epicurus and Gassendi, which claims that time is universal and infinite, is refuted by the same arguments whereby we showed that it is contradictory to claim that space is absolute. And how much Clark and Newton fall away from the truth in confusing time and eternity is evident from what we said concerning eternity and time. It is also clear that time is not an *a priori* perception of internal sensation, as Kant claims. For we proved that time presupposes a foundation in reality. Therefore, it is not something purely subjective. – Further, many sciences deal with time; for astronomy deals with the measuring of time, and geology makes arguments based on the antiquity or newness of geological strata. But science inquires concerning realities, not concerning merely interior forms. Therefore, time is not a purely subjective form.

Finally, many arguments show that time is not the order of successions, as Leibnitz thought.

1st Argument. Time or successive duration is not obtained from successive duration itself. But order results from successive duration; for there is an order in the succession of things due to the fact that there is a motion according to which things occur succession. It is not the case that there is motion because there is order. Therefore, time is not the order of successions.

2nd ***Argument***. Time requires continuous motion. But order can be without continuous motion, for there is an order of succession among the thoughts of angels, but there is no continuous motion. Therefore, unless we want to confuse our time with that of angels, we must acknowledge that time is not the order of successions.

3rd ***Argument***. Time, like any duration, requires the permanence of the enduring thing. But the order of succession can be without the permanence of the enduring thing; for indeed an excellent order is saved between two, three, or more entities that are in succession, even though none of them endures. *Therefore.*

SEVENTH ARTICLE

On the qualities of bodies[123]

I. – The Quality of Bodies. There are two genera of properties that are ascribed to corporeal substances: some result by reason of the matter, and others are by reason of the form. That which follows by reason of the matter is quantity, and what belongs to it by reason of the form is quality. We already discussed quantity and those things that are referred to it; now we add certain things concerning quality.

Now, quality is "an accident that comes by reason of the form and modifies the substance in itself."

It is said to be (1) an *accident*, that is, a predicamental accident or property,[124] which is not the essence of the thing but presupposes it and adds to it *esse secundum quid.*

It is said to be (2) *by reason of the form*, because form is the root or the cause from which it emanates; but yet, quality is not the form itself, nor does it have the form itself as its subject, just as matter is the root of quantity, without, however, quantity being received immediately in matter. Further, the first subject of quantity is the composite, while its proximate subject is quantity, which is first received in the substance and renders it determinable and open to act through other accidents.[125]

It is said that it (3) *modifies the substance in itself.* For this reason it is distinguished from: (a) quantity, which does not properly modify the substance but extends it into parts and renders it modifiable through other accidents, and (b) from the other accidents, which, of course, affect the substance, not in itself, but rather in relation to something extrinsic.

– In Logic we numbered four species of quality: *habit* and *disposition*, *potency* and *impotency*, *form* and *figure*, *passion* and

---

[123] One may consult Pesch, *Phil. nat.* nn. 360ff; Mielle, *Substantia corporalis*, pp. 302ff; Bulliot, "L'unité des forces physiques," *Annal. de Phil. chrét.* May-June 1889; Nys, *Le problème cosmologique* et *Cosmologie*; de Munnynck, "Les propriétés des corps bruts," *Revue Thomiste* 8; Duhem, *Evolution de la mécanique*; etc.

[124] A '*predicamental* accident' is an 'accident' understood as that which exists in a substance, as opposed to a '*predicable* accident', the last of the five predicables, a property that neither denotes nor is derived from the essence of the thing. – ***The Translator.***

[125] Cf. I. P. *Phil. nat.* Tract. II, q. 2, a. 3, n. 6.

passible quality. Now, in a body there are no habits properly speaking, but at most habitual dispositions that make the subject apt in relation to the form, such as *health*, *beauty*, etc. It remains, therefore, that we inquire concerning the other three species.

II. – PHYSICAL POWERS OR POTENCIES. In the present discussion, 'potency' is not taken generally insofar as it is opposed to act, but insofar as it designates a proximate and immediate principle of operation to which its own nature is inclined.[126] Taken in this way, in relation to operation, potencies are commonly called 'physical powers' by more recent authors. But what is the nature of such powers? The dynamists, who claim that substance is immediately operative, confuse the powers with the substance itself. But according to the atomists, all the activities of bodies are reduced to locomotion, and they do not acknowledge any principle of action besides matter and motion. And finally, the Scholastic doctrine can be expressed thus:

III. – CONCLUSION: "Physical powers cannot at all be reduced to matter and motion; rather, there are intrinsic principles that are really distinct from both the substance and the substantial form which nonetheless emanate from the form and are rooted in the substance."

We assume, against the occasionalists and idealists, that bodies possess a true activity; otherwise there would not be real motion in the world, real evolution, or any production of new substances.[127] Assuming this, we prove the thesis in parts.

(1) *Physical powers cannot be reduced to matter.* The principle of inertia and of potentiality cannot be the principle of activity and of perfection; for indeed, inertia and activity, potentiality and perfection are complete opposites and altogether irreducible. But matter is the principle of potentiality and inertia, whereas a power, on the other hand, is a principle of activity and perfection. Therefore, powers cannot be reduced to matter.

– (2) *Nor can they be reduced to pure motion.* Motion is the act of something that exists in potency insofar as it is in potency, or something

[126] We shall have a more detailed discussion concerning the nature and division of faculties in IIa-IIae, where we discuss the powers of the soul.

[127] We shall discuss the activity of bodies in *Ontology*, in the treatise on causes.

that is always in the process of becoming. Further, an *ens* that is in the process of becoming can exist and last only if some cause or motive power perseveres which puts the mobile thing in motion. Therefore, motion neither occurs nor perseveres without a certain motive power that puts the mobile thing in motion. But a motive power is a physical power. Therefore, the power presupposes motion; therefore, the power is not reducible to motion but rather is distinguished from it as a cause from its effect.

Further, physical powers are specifically distinct, as we have shown *passim*.[128] But this specific distinction cannot come from motion, which only implies a distinction *secundum quid* and in degrees. Therefore, physical powers are not reducible to motion.

Although physical powers are not motion, nonetheless they are not without motion. For, since they have quantity as their [immediate] subject, the modification that affects them indirectly changes quantity; and, because quantity implies parts in place, hence locomotion necessarily follows.

This is also evidenced by science, which has established as a law that the exercise of every physical power is concomitant with locomotion.

Thus to physical powers we must ascribe both something common and generic, namely that they produce motion, and something proper and dynamic by reason of which they are distinguished specifically.

– (3) *Physical powers are intrinsic principles.* From the consideration of locomotion alone it is altogether obvious that these powers are not God Himself or the divine will.

It remains, therefore, that they are principles inherent to the mobile thing itself. Further, bodies have a stable and firm mode of acting that varies according to the various bodies that exist. But a stable, firm mode of acting proper to each body argues for a principle of acting that is firm, stable, and proper, and thus intrinsic. Therefore, physical powers are intrinsic principles.

– (4) *They are really distinct from substance and substantial form.* The principle of operation must be in the same genus in which operation is, as we have noted many times, and as we shall prove in

[128] The Scholastic doctrine on the specific distinction of physical powers seems to be confirmed by scientific facts. Cf. NYS, *Cosmologie*, n. 226, and DUHEM, *Evolution de la mécanique*, IIe partie.

Metaphysics;[129] for indeed, act and potency are in the same genus. But operation is in the genus of an accident [i.e., action]. Therefore, a power or operative principle is not in the genus of substance; therefore, it is neither a substance nor a substantial form, but is properly speaking an accident.

– (5) *They emanate from form.* Secondary active principles must emanate from the first active principle. But powers are secondary active principles, and form is the first active principle, since form is the first act and the first principle of action. Therefore, powers emanate from substantial form.

– (6) *Has the substance as its subject.* We have already proven[130] that the first subject of accidents is not matter or form, but the composite, which immediately receives quantity and, through quantity, other accidents.

IV. – How Many Physical Powers There Are. They are reduced to three more general classes: *static powers*, *conservative powers*, and *communicative powers*. Static powers include cohesion, expansion, resistance, elasticity, repulsion, etc.; *conservative* powers include inertia and reaction; and *communicative* powers include chemical affinity, attraction, impulse, etc.

V. – Form and Figure. Although form and figure differ accidentally, in the present discussion they are taken as interchangeable. Now, figure is defined by St. Thomas as: "The form of a *quantum* [i.e., of a thing that has quantity] insofar as it is a *quantum*."[131] For indeed, the form of a thing, insofar as it is a substance and a body is said to be a substantial form; but the form of a thing, reduplicatively insofar as it is a thing that is a *quantum*, is called figure. Hence, figure is very closely related to quantity.

But in a stricter sense it is defined as: "A quality, or mode, that results in a body from the termination of quantity" (*Qualitas, seu modus, resultans in corpore ex terminatione quantitatis*). It is distinguished into *intrinsic* and *extrinsic* figure. Intrinsic figure is the terminus of quantity in relation to itself, e.g., the figure of the hand in itself; but extrinsic

---

[129] *Ontology*, Treatise I, q. 2, a. 3.

[130] Q. 2, art. 3, n. 7 of the present Treatise.

[131] *Summa theologiae* Ia, q. 7, a. 3: "[F]orma quanti, in quantum hujusmodi."

figure, or local or positional figure, is the terminus of quantity in relation to a place or position; thus, the hand, when open or closed, exhibits diverse positional figure. The former belongs to the essence of figure, whereas the latter, since it is a secondary effect, can vary without the former varying; indeed, they can be separated by divine intervention, just as secondary effects are separated miraculously from their causes. In the Eucharist, Christ has figure in the first way because His quantity has a terminus in relation to itself; but it does not have positional figure since diverse parts of quantity exist in the same place.

VI. – How Figure is Related to Substantial Form. Figure emanates from substantial form prior to the other qualities.

Proof of the First Part. Just as matter is indifferent to the formation of this or that substance, so it remains indifferent to this or that terminus of quantity. Therefore, the same principle that determines matter to a certain substance also determines the quantity of matter to a certain terminus, or figure. But the principle that defines matter to a certain substance is substantial form. Therefore, the determination of the quantity of matter to a certain figure also comes from the substantial form.

Proof of the Second Part. That quality approaches substantial form the most which approaches quantity most immediately and prior to all other qualities; the reason is that quantity is the first accident of a substance. But figure follows quantity more immediately than and prior to all other qualities because it is its form, or terminus. Therefore, figure is more closely related to substantial form than the other qualities.

This is confirmed by the authority of St. Thomas. "Just as, among accidents, quantity is the one that is most closely related to substance, so figure, which is a quality of quantity, is most closely related to substantial forms [than other accidents]."[132]

VII. – Figure and Species. Since figure is proximately accompanied by form and the form is what gives the species to the thing, figure is generally cited as an *accident characteristic of the species*. Hence, from the diversity of figure one is often able to infer the diversity of species: thus, in plants and in animals, a certain sign of the diversity

[132] *In VII Phys*, lect. 5: "Sicut quantitas propinquissime se habet ad substantiam inter alia accidentia, ita figura, quae est qualitas circa quantitatem, propinquissime se habet ad formam substantiae."

of species is a diversity of figure. Similarly, in crystals, a diversity of specific types can be deduced from their diversity in figure. But the figure itself of the crystal is not reducible to [the extrinsic forces of] attraction alone, which, since they are the same, cannot constitute the six marvelous types of crystals; rather, the crystal exhibits a higher principle, which is the substantial form.[133]

VIII. – Sensible Qualities. Sensible qualities are those which are apprehended by the senses or that generate a bodily change in the senses. They are divided into primary and secondary sensible qualities. Primary sensible qualities are defined by the ancients as "those which are derived neither from themselves nor from others, but others from them" (*quae nec ex seipsis nec ex aliis fiunt, sed aliae ex ipsis*). They posited four: *heat* and *cold*, *dryness* and *humidity*. Secondary sensible qualities were those which result from various degrees of mixing and mutual moderation of the primary qualities. But primary and secondary sensible qualities were usurped by Locke and his followers in a different sense. For them, primary qualities are common sensibles, which are perceived by different senses, such as *extension* and *magnitude*, whereas secondary qualities are proper sensibles, such as *color*, *odor*, etc. Recent authors use the term 'primary quality' to refer to those qualities that are irreducible, that is to say, which are not reducible to other, simpler qualities; just as simple bodies are those which are not resolved into others.[134]

There is a bitter disputation on the objective reality of sensible qualities. Locke admitted the reality of primary qualities but altogether rejected that of secondary qualities; and the Cartesians, the Kantians, and many recent authors, both philosophers and physiologists who claim that qualities are nothing other than motions and vibrations [in the nervous system], deny the objectivity of qualities in general.

IX. – Conclusion: "All sensible qualities really exist on the part of things, just as they are apprehended by the senses."

1st Argument. Experience and reason attest that there are specifically diverse sensations in us because the eyes perceive diverse species of color and the ears receives other specifically diverse

[133] See above, q. 1, a. 4, n. 4, arg. 3.

[134] Cf. P. Duhem, *Evolution de la mécanique*, II P., ch. 1.

impressions, etc. But this diversity of sensations cannot be explained unless sensible qualities bear an objective reality. *Therefore. Proof of the Minor.* A species argues for some formal principle that is constant and permanent. But such a principle cannot be motion and vibrations, but it is necessary that it be something objective. For diversity of motions is only diversity in degrees. But a difference in degrees does not cause specific diversity, as is clear that from the different degrees of heat there does not arise a new species of heat. Therefore, the specific diversity in sensations cannot be explained if qualities are reducible to vibrations and motion.

2nd Argument. The cognition whereby the senses receive sensible qualities is an immediate and intuitive cognition. But every cognition, and especially intuitive cognition, is an assimilation of the faculty with its object, for indeed the function of cognition is not to produce its own effect but to express the object vitally and to assimilate itself to it. Therefore, sensible qualities and the senses are formally assimilated. Therefore, sensible qualities exhibit an objective reality, just as they are apprehended by the senses.

Confirmation. Since the senses by their nature are ordered to sensible qualities as to their common or proper sensibles, unless qualities exist as they are apprehended we would have to admit that the senses err concerning their object, which is unacceptable. The thesis, therefore, is established from all those things whereby we vindicated the veracity of the external senses and refuted the arguments of the adversaries.[135]

Nor is it only the Scholastics, but also many scientists, who profess the reality of quality and its being distinct from quantity:

> "Hence we are obliged to accept in our Physics something other than the purely quantitative elements of which geometry treats, to admit that matter has *qualities*: at the risk of being interpreted as defending a return to *occult powers*, we are forced to regard as a primary and irreducible quality that by which a body is hot, or lit, or electrified, or magnetized; in a word, renouncing the attempts that have been unceasingly begun since Descartes, we need to refasten our theories to the most essential notions of peripatetic Physics."[136]

---

135 *Major Logic*, Treatise II, q. 3, a. 2.

136 P. DUHEM, *Evolution de la mécanique*, p. 197-198: "Nous voici donc obligés de recevoir en notre Physique autre chose que les éléments purement

FOURTH QUESTION

On the principle of individuation[137]

Having studied what body is with respect to its essence and with respect to its properties, another somewhat difficult question arises: how corporeal substance is individuated.

Because the solution to this question depends on the notion of matter and quantity, it is more fitting to discuss the matter here, after having discussed matter and quantity. The question really concerns the individuation of corporeal substances; but for the sake of the connection of this doctrine with others, we add a few things concerning the individuation of the human soul and of accidents.

FIRST ARTICLE

On the individuation of bodies

I. – The State of the Question is Explained. In things that are composed of matter and form we see many things that coincide in species but differ in number; the whole and entire *ratio* of a species is in each, and nonetheless, whereas the individuals are multiplied, the

quantitatifs dont traite le géomètre, d'admettre que la matière a des *qualités*; au risque de nous entendre reprocher le retour aux *vertus occultes*, nous sommes contraints de regarder comme une qualité première et irréductible ce par quoi un corps est chaud, ou éclairé, ou électrisé, ou aimanté; en un mot, renonçant aux tentatives sans cesse renouvelées depuis Descartes, il nous faut rattacher nos théories aux notions les plus essentielles de la Physique péripatéticienne."

[137] On the principle of individuation, see Aristotle, *Metaphysics* 11.8; St. Thomas, *Summa theologiae* Ia, q. 50; *Super Boethium De Trinitate* q. 4, a. 2; *De ente et essentia*, c. 2; the opusculum *De principio individuationis*; the opusculum *De natura materiae*, c. 3; *Quodlibet* II, q. 2, a. 4; Scotus, *In II Sent.* d. 3, q. 1; Capreolus, *In II Sent.* d. 3, q. 1; Cajetan, *In De ente*, c. 2, q. 5; Suárez, *Disput. Metaphys.*, disp. 5; John of St. Thomas, *Phil. nat.* q. 9, aa. 3-5; Salmanticenses, tom. I, *De principio individuat.*; De Maria, *Ont.* Tract. 2, q. 3, a. 5; Mercier, *Ont.* nn. 36ff; Nys, *Cosmologie*, nn. 211ff.

species is not thereby multiplied. It is asked, therefore, how it is so that there are many substances without the species being multiplied such that there are many individuals that are substantially distinct but nonetheless coincide in the same nature. This is the celebrated and difficult question on the principle of individuation which seemed irresolvable to Bossuet.[138] And it is of great importance since it is, above all, convenient to discern the *ratio* and the properties of individuals from the *ratio* and properties of the species.

The individual is the primary substance itself, which the Philosopher calls *this something* (*hoc aliquid*, *tode ti*). Now, four things pertain to the *ratio* of an individual. It is required that it: (1) be undivided in itself; (2) be divided from others, if there be any individuals that exist or are able to exist in the same species; (3) not be communicable in the same way in which a universal is attributed to singulars; (4) be something demonstrable and determinate to here and now.

And so the principle of individuation is that whereby nature is first rendered incommunicable and whereby it is distinguished from others of the same species. Now, individuals are distinguished intrinsically and substantially from each other. Therefore, that whereby they are first distinguished must be an intrinsic and substantial principle. For this reason the principle of individuation can be defined as: "A principle, not efficient but intrinsic and substantial, whereby a substance is *FIRST* multiplied without the species being multiplied; whereby nature is *FIRST* rendered incommunicable, and whereby it is *FIRST* distinguished from others of the same species" (*Principium, non quidem efficiens sed intrinsecum et substantiale, quo* PRIMO *fit ut multiplicetur substantia non multiplicata specie; quo* PRIMO *natura redditur incommunicabilis, et* PRIMO *distinguitur a caeteris ejusdem speciei*).

Hence, we are dealing, not with an immediate principle but with the *first* root from which individuation arises and is derived.

II. – OPINIONS ARE PRESENTED. There are many opinions on this matter. (1) Aureolus, Durandus, Gregory of Rimini, Gabriel Biel, and other nominalists—with whom are in agreement Suárez and other more recent authors, such as Rosmini, Gioberti—hold that material substances are singular and individual of themselves; and so they posit the entity

---

[138] Cf. *Logique*, L. I, c. 33.

itself as the principle of individuation. The second opinion is that of Scotus: the Subtle Doctor thinks that nature is individuated through its 'thisness' (*haecceitas*), some positive mode, which is neither matter nor form nor the composite, but a mode of all of them, or the ultimate reality of matter, form, and the composite, which comes extrinsically to the nature and renders the nature indivisible into subjective parts. Hence, Scotus reaches this conclusion: "The principle of individuation is neither matter nor form nor quantity, but an individual property, namely, 'thisness'." (3) There were also some who assigned subsistence and existence as the principle of individuation, as Scotus relates.[139] (4) Avicenna, Averroes, Zabarella, and some Scholastics believe the principle of individuation is the substantial form. (5) The solemn position of the Thomists is that the first principle of individuation is matter designated by quantity (*materia signata quantitate*).

III. – FIRST CONCLUSION: "In matter-form composites the quiddity is not singular and individual *per se*."

1st Argument. If the quiddity were essentially and *per se* individual, it would imply individual unity *per se* and essentially. But the quiddity of a material thing requires neither plurality nor unity *per se* and essentially. Therefore, the quiddity of a material thing is not *per se* and essentially individual. *Proof of the Minor.* From the words of St. Thomas:

> "For if plurality belonged to its concept [i.e., of nature], it could never be one, although nonetheless it is one according as it is in Socrates; similarly, if unity belonged to its *ratio*, then [the nature] of Socrates and of Plato would be one and the same, and it could not be multiplied in many [individuals]."[140]

---

[139] FENELON (*Traité de l'existence de Dieu*, IIe partie, ch. 4) also holds this view: "If one wishes, in good faith, to consider actual existence without abstraction, it is true to say that it is precisely that which distinguishes one thing from another... This produced existence is the singular being or the individual" (*Que si on veut, de bonne foi, considérer l'existence actuelle sans abstraction, il est vrai de dire qu'elle est précisément ce que distingue une chose d'une autre... Cette existence produite est l'être singulier ou l'individu*).

[140] *De ente et essentia*, c. 4: "Si enim pluralitas esset de intellectu eius, nunquam posset esse una, cum tamen una sit secundum quod est in Socrate.

See also what we said concerning this issue in *Major Logic*, Treatise I, a. 1.

2nd Argument. If quiddity were individual *per se*, it would be interchangeable with the individual, and consequently it could be predicated of the individual in the abstract. But in material things the essence is not predicated of the individual in the abstract; for one does not say, "Socrates is his humanity," but rather, "Socrates is a man." Therefore, the essence in material things is not individual *per se*.[141]

Objection. There is a distinction of reason between an essence and an individual, and similarly, there is distinction of reason between the concrete and the abstract, and hence it is that the essence in the abstract cannot be predicated of the individual in the concrete. ***Reply***. The distinction between God and the Deity is one of reason, and nonetheless, the following proposition is true: "God is the Deity." Therefore, the distinction of reason between an essence and an individual cannot prevent the truth of the proposition "Peter is his humanity." Therefore, the distinction between the essence and the individual is not a distinction of reason, but a real distinction; therefore, the essence is not *per se* individual.

IV. – SECOND CONCLUSION: "A '*thisness*' (*haecceitas*) that comes extrinsically cannot be the principle of individuation."

Thisness can be taken in two senses: (1) as being the same as the individual difference, which formally makes a nature to be a singular *this*; (2) as being a mode that comes extrinsically. If Scotus only meant that nature formally becomes individual through an individual difference, then he spoke the truth, but missed the point of the difficulty. For we certainly admit that there is an individual difference, but we inquire what is the root and principle of this individual difference. – But if he argues that the principle of individuation is some mode that comes extrinsically, then he wandered off very far from the truth.

For one can argue against him thus: *Thisness* understood in this way is something either accidental or substantial. But neither can be

---

Similiter si unitas esset de ratione eius, tunc esset una et eadem Socratis et Platonis nec posset in pluribus plurificari."

[141] Concerning this opinion, D. NYS rightly concludes that: "Such a solution is only acceptable in the absence of any other, for, in fact, it explains nothing" (*Pareille solution n'est acceptable qu'à défaut de toute autre, car, en fait, elle n'explique rien*).

admitted. *Therefore. Proof of the Minor with Respect to the First Part.* That which constitutes a primary substance or a substantial individual cannot be an accident. But *thisness* is supposed to constitute the substantial individual. Therefore, it is not an accident. *With Respect to the Second Part.* If it is something substantial, then it is either a substantial entity or a mode of a substance; if it is the entity itself, then we have already fallen into the opinion of the nominalists, which asserts that the entity itself of the thing is the principle of individuation; but if it is the mode of a substance, then it does not individuate but presupposes an already constituted substance and is founded in it, for a mode presupposes a thing that it modifies.

V. – THIRD CONCLUSION: "The principle of numeric individuation is not existence or subsistence."

This hardly needs proof. Existence and subsistence are really distinct from the essence because they come (*adveniunt*) to an essence, and so they presuppose an already constituted and individuated essence. But if someone were to argue that existence is not really distinct from essence, but that it is the essence itself, we would further have to ask through what the essence and existence itself is individuated.

VI. – FOURTH CONCLUSION: "The principle of individuation is not a form."

The principle of individuation must be an incommunicable root, as is evident from what we have said. But form is rather the principle of communicability. *Therefore. Proof of the Minor.* The form is the principle of specific degree. But a specific degree is of itself communicable to inferiors. Therefore, the form is the principle of communicability.

– ***Another argument***: A specific difference is opposed to a numeric difference. But form is the principle of specific difference. Therefore, it is not the principle of numeric difference.

VII. – FIFTH CONCLUSION: "The principle of individuation is neither matter alone nor quantity alone."

The principle of individuation is that whereby nature becomes determinate and incommunicable. But matter according to itself is indeterminate and indifferent to being in this or that individual. Therefore, matter alone is not the root of individuation. – Nor is quantity. For indeed, the principle of individuation is something substantial. But quantity of itself is an accident. *Therefore.*

VIII. – Signate Matter. – Various Explanations. – Thomists unanimously teach that the principle of individuation includes both matter and quantity, but matter's being signate through quantity is explained in various ways. Some say that signate matter is a certain aggregate of matter and actual or informing quantity, and they claim that this aggregate is the principle of individuation. Two things, they say, belong to the *ratio* of the individual: incommunicability and distinction from others; and matter provides incommunicability, whereas quantity provides distinction from others. This is the view of Ferrara, Soncinas, and Nazarius. But this opinion is not satisfactory. Matter's being signate must inhere in the matter itself. But informing quantity inheres, not in matter, but in the whole composite. Therefore, matter does not become signate through informing quantity. Further, informing quantity, just like the other accidents, gives only accidental *esse*. But accidental *esse* cannot essentially constitute the individual. Therefore, informing quantity does not make matter signate. – Others defend the view that matter becomes signate through a certain substantial mode that comes from the nature of the thing and that is distinct from the matter and which determines it to this rather than that quantity. – But they posit this mode needlessly. Moreover, this mode is either individuated by matter or is individual *per se*. If it is individuated by matter, then it is already presupposed that matter, without this mode, is the principle of individuation at least with respect to the mode; and for the same reason it would be the principle of individuation with respect to everything else. But if it is singular *per se*, then something is individuated *per se* without matter. Why are not the other modes and forms and accidents this way?[142]

IX. – The True Notion of Signate Matter. The true explanation, therefore, is the one that Cajetan, Bañez, John of St. Thomas, and the Salmanticenses give, namely: matter designated by quantity is prime matter itself, as precontaining in the manner of a root this quantity rather than that, or matter capable of having this quantity,

[142] Cf. John of St. Thomas, *Phil. nat.* II P., q. 9, a. 4.

such that it is not capable of having that quantity.[143] Just as a thing marked with a seal comes to belong to someone in such a way that it cannot belong to another, so signate matter, or 'sealed' matter, is capable of having this quantity in such a way that it cannot be of another quantity.

Signate matter, therefore, means directly matter itself but indirectly the quantity that it connotes, just as the potency of matter directly signifies matter but indirectly the act that it receives and by the *ratio* whereby it is defined. Therefore, the issue is to be understood thus: matter is individuated *intrinsically* through a connotation of such a quantity, but *extrinsically* through the connoting quantity itself; in other words, *the relationship to this quantity* individuates in the manner of an intrinsic principle; but *this quantity* individuates in the manner of the extrinsic things connoted. Form is individuated by matter thus individuated, and the whole entity of the thing—from which the *thisness*, or the individual difference, is immediately derived—is finally individuated by the form. Therefore, signate matter is not said to be the principle of individuation in the sense that it is the individual difference itself, but because it is that from which the individual difference is first derived. Just as the intellectual soul is not the difference of man itself, namely, 'rational', but that from which it is derived, so signate matter is not the individual difference, but that which intrinsically causes it. And, just as a man is man both through its 'rational' as through its total formal principle and through the intellectual soul as through its partial formal principle by which 'rational' is caused, so an individual substance is constituted as *this* both through the individual difference as through its total principle and through signate matter as through its prior intrinsic cause by which that individual difference itself is caused.[144]

---

[143] NYS, *Cosmologie*, n. 214: "The capacity of matter with regard to this quantity is not an adventitious property distinct from matter: rather, it is identical to it; in other words, it is matter itself, in its relation to a reality that is to come" (*La capacité de la matière à l'égard de telle quantité n'est pas une proprieté adventice distincte de la matière: elle s'identifie au contraire avec elle; en d'autres termes, elle est la matiere elle même, affectée par nous d'une relation avec une réalité à venir*).

[144] Cf. CAJETAN, *In De ente*, c. 2, q. 5.

X. – Sixth Conclusion: "Signate matter, in the sense explained, is the first principle of numeric individuation."

Proof. The principle of individuation must be substantial, intrinsic, and the principle of both incommunicability and of distinction from others. But signate matter exhibits all these things. *Therefore. Proof of the Minor.* (1) Matter in itself is an intrinsic and substantial principle and does not cease to be substantial by the fact that it becomes signate; for becoming signate is not an accident, but a transcendental relation to quantity. (2) It is a principle of incommunicability. For indeed, signate matter is capable of having this quantity in such a way that it is not capable of having another; indeed, it is not intelligible under some other quantity. But, if it is not intelligible under another quantity, it is impossible for it to be communicated to another which has some other quantity. Therefore, signate matter is incommunicable. (3) Signate matter, or matter that has a relation to this quantity, is the principle of distinction from others.

For it belongs to the *ratio* of quantity to have parts outside of parts or for one part not to be another. Therefore, two parts of quantity are in virtue of their essence distinct from each other, and similarly two quantities are *per se* distinct by their position and place. But that which is *per se* such is the cause of others. Therefore, quantity is the cause of the distinction of others. Therefore, matter that receives quantity *B* will be intrinsically distinct from matter that is related to quantity *C*. Consequently, a form that is received in matter that has quantity *B* will be intrinsically distinct from a form that is received in matter that has a relation to quantity *C*; but a distinction of forms implies a distinction of composites. Therefore, signate matter, or matter that has a relation to a determinate quantity, is the principle of distinction from others.

We have established, then, that the truth is as we asserted, namely, that matter becomes individual through its relation to quantity; that form is individuated by matter thus individuated and signate; and that the composite is individuated proximately by the form.

XI. – A Difficulty is Handled, and it is Explained How Matter Receives This Quantity Before This Form.

We said that matter is individuated because it has a relation to quantity, and if this were established, the thesis would remain unshaken.

But an objection arises immediately: Matter is first (*prius*) related to that which is prior (*prius*) in the subject. But form is prior to quantity in the subject. Therefore, matter is first related to this numerically [one] form than to this numerically [one] quantity, and so it is not individuated by quantity but by form. ***Reply***. *I distinguish the major:* that matter is first related to that which is prior in the thing in the *ratio* of dispositive and determinative cause, I concede; but that it is first related to that which is prior in the thing in the *ratio* of a purely dispositive cause, I deny. *I contradistinguish the minor*: that form is prior to quantity in the *ratio* of purely receptive cause, I concede; but that it is prior in the *ratio* of dispositive and determinative cause, I deny. *And I deny the conclusion.*

Explanation of the Solution. Form and quantity precede each other in diverse orders of causality. For in the genus of purely receptive cause, the substantial form is prior, since it is received in matter immediately, whereas the quantity is received mediately and after the reception of the form. But in the genus of dispositive cause, quantity is prior. For matter is related to this rather than that form only through dispositions which determine the potentiality of matter with respect to one form over another. For unless these dispositions coincided, matter would remain altogether indifferent. Therefore, matter is first related to the dispositions that determine it to a certain form than to the form itself, which is cause in a certain way by the dispositions and thus is posterior to them. But the first and most determinative of all dispositions is quantity. Therefore, matter is first related to quantity than form. *Proof of the Minor*. Quantity is the first accident, and it belongs to its *ratio* that it divides one part of matter from another, and when this division has been made, there result a distinction and a determination with respect to the matter thus divided; but matter thus divided and determined becomes capable of having such determined form. Therefore, quantity is that which most disposes and determines matter with respect to form. It is clear, therefore, that quantity is prior to form in the genus of dispositive cause.

Nor is it inconvenient that the same thing precedes and follows form under different respects. For causes, as the axiom says, are causes of each other (*causae... ad invicem sunt causae*): thus the opening of the window—to use a trite example—under one *ratio* is the effect of the blowing wind, and is posterior to the blowing; under another respect it is its cause and it is prior to it; for the wind cannot enter unless the window is open. Therefore, this view rightly conceives how quantity can be prior to form in the genus of dispositive cause and posterior to in the genus of receptive cause.

Now, it is easy to understand why matter receives that which is prior (*prius*) in the *ratio* of dispositive cause before (*prius*) it receives that which is prior (*prius*) in the *ratio* of receptive cause. Since a disposition to form is the *ratio* of receiving a form, to be disposed to form is prior to receiving a form. Therefore, matter must relate first (*prius*) to that which disposes and determines it to form than to the form itself that is to be received. Since, therefore, quantity is a dispositive cause, matter must relate first to quantity than to form, and hence it happens that it is individuated through the connotation of this quantity and not through the connotation of this form.

XII. – Another Difficulty is Resolved by Applying the Distinction of Limited Quantity and Unlimited Quantity.

It is hardly understandable how quantity can individuate. For indeed, quantity presupposes its subject. Therefore, it is posterior to the subject and, therefore, it does not individuate the subject but is rather individuated by it. ***Reply***. *I distinguish the antecedent*: that a *limited* quantity presupposes its subject, I concede; but that *unlimited* quantity presupposes a subject, I deny. *I thus distinguish the conclusion*: that, therefore, *limited* quantity is individuated by the subject, I concede; but that *unlimited* quantity is individuated by the subject, I deny.

Quantity can be seen in two ways: (1) as a perfect accident, as it inheres and informs, as it is perfected through the substantial form, which is its complement; (2) as it is an imperfect accident, abstracting from a certain terminus and figure. In the first case it presupposes a subject and is individuated by it; but thus taken, quantity is only a principle of individuation by extrinsic connotation. Seen in the second way, it does not presuppose a subject but is prior to it as a root that is pre-contained in the matter of individuation of that subject. Quantity in this respect is individual *per se*. For indeed, the subject is what makes quantity actually inhere and inform; but in virtue of its essence, and independently of the subject, it requires having parts outside of parts, whereby it happens that two parts of quantity and two quantities are *per se* distinct. Therefore, quantity is distinct and individual *per se* and in virtue of its essence.

– ***But one could further insist on this objection***: Quantity cannot determine and individuate unless it itself is determinate. But it is not *per* *se* determinate, but only through the form. Therefore, form, not quantity, is the principle of individuation. ***Reply***. *I distinguish the major*: that quantity cannot individuate unless it itself be determinate, in the sense that it refers to *this* numerically [one] quantity rather than *that*, I concede; but that it cannot individuate unless it itself be determinate by a last determination, such that it has a certain terminus and figure, I deny. *I contradistinguish the minor*: that quantity cannot be determined except through form by a last and perfect determination, I concede; but that it cannot be determined except through form by a determination present in it such that it refers to *this* numerically [one] quantity rather than *that*, I deny. *And I deny the conclusion.*

Explanation of the Solution. The Salmanticenses say: "*Determinate* (*determinata*) quantity is distinct from *limited* (*terminata*) quantity. For determinate quantity means quantity as numerically *this* quantity and not that, not caring whether it is limited and complete through substantial form or whether it has this or that magnitude, and in this way it is the principle of individuation, insofar as it is pre-contained in matter. But limited quantity means the same quantity as being posterior in nature to the substantial form, through which it is limited and formed; and thus it implies according to that *ratio* that it has a certain magnitude, e.g., two cubits long; and taken in this sense it is not the principle of individuation."[145]

Therefore, in order for quantity to be able to individuate, an ultimate and complete limitation is not required; it is sufficient that it imply numerically *this* rather than that quantity. For two quantities, since they are *per se* distinct, divide *per se* one portion of matter from another; but after that division is done, there results, as we said, a distinction with respect to matter thus divided, and from matter distinguished that way there results a distinction with respect to the form. Therefore, for quantity to be the principle of individuation, it is

[145] *De princip. individ.* Tract. I, disp. 2, dub. 5, n. 132: "Aliud est quantitas *determinata*, aliud vero *terminata*. Quantitas enim determinata dicit quantitatem ut *hanc* numero et non illam, non curando an sit terminata et completa per formam substantialem, nec an habeat hanc tantitatem vel illam, et sic est principium individuationis, quatenus in materia praecontenta. Quantitas vero terminata dicit eamdem ut posteriorem natura quam forma substantialis, per quam terminatur et formatur; et similiter involvit secundum hanc rationem quod habeat certam tantitatem, v.g., bicubitam; et sic sumpta non est principium individuationis."

sufficient that the determination that is present in it imply numerically *this* rather than that quantity. Now, this determination does not come from the substantial form; for we have already shown how, thanks to the preceding dispositions, matter is first related to this numerically [one] quantity than to this numerically [one] form. Therefore, this numerically one quantity is contained radically in matter before this numerically one form, and hence the determination to this numerically one quantity is prior to the determination of the substantial form.

The said distinction between limited quantity and unlimited quantity is especially to be kept in mind. For, if by 'matter designated by quantity' one understands matter that has actually complete quantity and that is actually inhering in the manner of an intrinsic and informing cause, this 'matter designated by quantity' is no longer the principle of individuation with respect to its *esse*, but with respect to its *being known and being demonstrated* (*quoad cognosci et demonstrari*), as the Salmanticenses warn, insofar as it is its inseparable sign. But if 'signate matter' means matter itself as pre-containing this quantity rather than that in the manner of a root and foundation, then it is the principle of individuation also with respect to *esse*, as our arguments show.

XIII. – Last Difficulty. An objection that many great authors make now topples, given that we have employed the distinction above. The individual remains the same and the quantity varies; in fact, God can remove the quantity while the individual remains. But the principle of individuation is invariable. Therefore, it is not quantity. ***Reply***. The quantity that varies, which can be removed by God, is limited quantity, quantity as informing and as inhering. But, even if the individual remains, there remains in the matter the same, constant determination to *this* numerically [one] quantity over some other; hence the kind of quantity that is the principle of individuation, namely, quantity as pre-possessed and pre-contained in matter in the manner of a root, is invariable and inseparable from matter.[146] Therefore, *I distinguish the*

[146] Nys, *Cosmologie*, n. 215: "Nevertheless, under ordinary circumstances, each species of body has natural dimensions that distinguish it from others, and what is in question here is precisely this normal volume, whose variations are contained between two extremes" (*Néanmoins, dans les circonstances ordinaires, chaque espèce de corps a des dimensions naturelles qui le distinguent des autres, et c'est justement ce volume normal dont les variations sont comprises entre deux extrêmes assez rapprochés, qui est ici en question*).

conclusion: that informing and limited quantity is not the principle of individuation, I concede; but that unlimited and pre-contained matter is not the principle of individuation, I deny.[147]

[147] Our doctrine is summarized in the 11th Thomistic Thesis, approved by the Sacred Congregation of Studies: "Matter designated by quantity is the principle of individuation, that is, of the numeric distinction of one individual from another within the same specific nature—[individuation] which cannot occur among pure spirits" (*Quantitate signata materia principium est individuationis, id est, numericae distinctionis, quae in puris spiritibus esse non potest, unius individui ab alio in eadem natura specifica*).

Application of the aforesaid doctrine to the individuation of the human soul and to the individuation of accidents

I. – The Principle of Individuation in Man. We can examine three things in man: the body, the soul, and the composite. The body, or the matter, is individuated because it is related to this numerically [one] quantity; the soul becomes individual because it has a transcendental order to this body; and the composite, finally, is proximately individuated because it is informed by this soul. Even though these points can be somewhat established from the preceding discussion, nonetheless we shall now prove them again specifically, especially with respect to the soul. Hence, it is the case that:

II. – Conclusion: "The principle of individuation of the human soul is a transcendental order to the body."

Proof. The principle of individuation of the soul is that whereby this soul first differs from that one. But this soul differs from that because it has a transcendental order to this body, and that soul to that body. Therefore, the principle of individuation of the soul is the soul's transcendental order to this body. *Proof of the Minor.* The human soul is essentially the form of the body, such that the *esse* of the human body does not follow the human soul, but makes the soul human. But by the very fact that it is essentially the form of the human body, it essentially implies a commensuration or relation to the human body. Therefore, the commensuration of the soul to the human body does not follow a soul that is already human, but makes the soul human. Thus, *this* human soul is essentially the form of *this* body, such that its being the act of this body does not *follow* from this soul, but makes it *be* this soul. But the form of this body necessarily and essentially implies a commensuration, a *coaptatio* (its being made apt for) or relation to this body; for 'this act' signifies an essential relation to 'this potency'. Therefore, commensuration or relation to this body does not follow this soul, but makes it *this* soul. Therefore, the principle whereby the soul is *this* or *that* soul comes from the commensuration to this or that body; and for that reason the relation or *coaptatio* to this body is the principle of individuation of the intellectual soul. – This is so in man in a similar way to the way it is so in individuals in general, namely, the body or matter is individuated because it is related to this numerically [one] quantity, and the soul is made individual by the body thus individuated, and the whole individual human is individuated by the soul.

III. – The Individuation of the Soul, however, is not [Produced] by the Body as by an Efficient Cause. We must expressly note that the individuation of the soul does not depend on the body *causally*, but only *as an occasion* (*occasionaliter*). For 'form' is threefold: there is a form that is made *from* matter and *in* matter, as the form of a stone or the soul of an ox; another that is made neither *from* matter nor *in* matter, as the angel; and a third form that is *in* matter but is *not* made *from* matter or *out of* matter, namely, the human soul. Therefore, our soul does not receive being *causally* from matter.

Nonetheless, since the soul is *this* soul only due to the fact that it is the form of this body and that it implies a commensuration to this body, hence it follows that the body is a condition without which this soul attains neither *esse* nor unity. Therefore, the multiplication of bodies is an occasion *in becoming* (*occasio in fieri*) of the multitude of souls. But when an occasion in becoming ceases to be, the effect does not thereby cease to be with respect to its *esse*; hence when this body ceases to be, this soul does not thereby cease to be. For this reason the individuation of the soul by the body in no way prevents the immortality of the soul.

We conclude with a few words: The transcendental relation of the soul to this body is the soul's intrinsic principle of individuation, and this relation always remains in the soul, even after the destruction of the body, but this body is not its cause, properly speaking, but a necessary condition for the individuation of the intellectual soul, and if this condition ceases to be, the individuation does not thereby cease.

IV. – Difficulty Resolved. ***Objection***. The soul is not individuated by the body, but the body by the soul. For the body remains numerically the same in childhood, youth, and old age. But it does not receive its being numerically the same from quantity, which exists in perpetual flux and change. Therefore, it receives it from the soul, which is invariable. ***Reply***. *I distinguish the minor*: that it does not receive its being numerically one from actual and informing quantity, I concede; but that it does not receive its being numerically one from an essential relation to quantity, I deny. *And I deny the conclusion.*

The solution can be derived from what we discussed above. That which varies in the body is inhering and informing quantity; but the relation to this numerically [one] quantity rather than that other remains always and invariably.

The objection only shows that the body is not individuated through inhering quantity, which we already conceded of our own initiative; but in no way does it prove that the body is not individuated through an essential relation to quantity as radically pre-contained in matter.

Moreover, here an equivocation can easily creep in. For, by 'body' one can understand either matter itself or the composite itself.

Matter becomes individual directly through the connotation of quantity, whereas the human composite is individuated by signate matter as by a first principle and proximately and immediately by the soul itself. Hence, that that composite always remains numerically the same comes *proximately* from the soul; but one cannot thereby conclude that signate matter is not the *first* and *radical* principle of individuation.

V. – On the Individuation of Accidents. – Opinions. Durandus, Suárez, and many others hold that an accident is individuated *per se* insofar as it such and such an entity in act or in aptitude. St. Thomas, the Salmanticenses, and the Thomists unanimously hold that accidents are individuated by the subject, which is, as it were, their matter.

VI. – Conclusion: "Besides unlimited quantity, which is *per se* individual, all accidents are individuated by their own subject."

We have sufficiently discussed quantity: unlimited quantity is individuated *per se* because in virtue of its essence it has parts outside of parts, but limited quantity is related to the subject like any of the other accidents concerning which is the present conclusion.

***Proof of the Thesis*. (1) *From the Authority of St. Thomas*:** "Accidents are individuated through their subject": we say 'this white' insofar as it is in this subject.[148] "But just as accidents have *esse* in the subject, so they receive unity and multitude from the subject."[149] – "We

[148] *Summa theologiae* Ia, q. 29, a. 1: "[A]ccidentia individuantur per subiectum."

[149] *Ibid.*, a. 3: "Accidentia autem, sicut esse habent in subiecto, ita ex subiecto suscipiunt unitatem et multitudinem."

must note that accidents *are not individuated through prime matter, but are individuated through their own subject*, which is *ens* in act."[150]

(2) ***From Reason.*** ***1st Argument.*** A numeric difference is a material difference just as a specific difference is a formal difference. But a material difference must come from matter just as a formal difference is taken from the form. Therefore, the numeric difference of accidents is traced back to their matter. But the matter of accidents is their subject. Therefore, the numeric difference of accidents originates from the subject—not taken materially, of course, since in the same subject taken materially there can be many accidents, but taken formally insofar as it underlies *the same action, of the same agent, and at the same time.*

2nd Argument. The principle of individuation must be the cause of incommunicability and distinction from others. But the principle of incommunicability and distinction is not traced back to the accident itself, but to its subject. Therefore, an accident is not individuated *per se* but through its subject. *Proof of the Minor.* Accidents, as such, are acts and forms. But a form of itself is communicable, unless [its being communicated is] prevented and limited. Therefore, the principle of incommunicability is not derived from the accident itself.

– But from the subject. For indeed, the subject taken formally is that which underlies *the same action of the same agent and at the same time.* But from the very fact that an accident is received in such a subject, it becomes incommunicable to another [subject] that underlies other actions, or [the same subject] at other times; and thus it is distinct from an accident that is received in another subject that underlies another action or another time. Therefore, the subject taken formally is the principle of incommunicability and distinction.

VII. – DIFFICULTIES RESOLVED. ***1st Objection.*** Signate matter is the principle of individuation of substance. Therefore, *a fortiori* it is the principle of individuation of accidents. ***Reply.*** *I deny the conclusion.* For

---

[150] THOMAS SUTTON (?), *De principio individuationis*, towards the end: "Sed advertendum est quod accidentia non individuantur per materiam primam, sed per subiectum proprium quod est ens actu individuantur, sicut et formae substantiales per materiam primam, quae est earum subiectum." In Hugon's time, this work was attributed to St. Thomas, although now it is considered as being of doubtful authenticity, and some attribute it to Thomas Sutton (d. 1315), an early Thomist. – ***The Translator.***

accidents are secondary forms and have *esse secundum quid*. Therefore, they require a subject that already has a first form and *esse simpliciter*, and this is the composite.

2nd Objection. A thing is made numerically one through that through which it is an *ens*. But an accident is made an *ens* through its own entity. Therefore, it is made one thing through its own entity; that is to say, it is individuated *per se*. ***Reply***. *I distinguish the major*: that a thing becomes numerically one through that through which it is made an *ens*, as through a proximate and immediate principle, I concede; but that a thing becomes numerically one through that through which it is made an *ens*, as through a first and radical principle, I deny. *I distinguish the conclusion*: that, therefore, an accident becomes numerically one through its own entity, as through a proximate and immediate principle, I concede; but that it does so as through a first, radical principle, I deny.

– In this whole question we are dealing with the first principle and root of individuation, whether we speak in reference to substance or to accidents. For, if we are discussing the proximate principle of individuation, we could concede that the composite is individuated *per se* through its individual difference and less remotely through the substantial form. But that in no way prevents signate matter from being the remote and radical principle of individuation. Thus, we acknowledge that accidents are individuated *proximately* through their entities insofar as each of these is a *this*, but we inquire the *first* cause of why they are *this*. We have shown, however, that these are *this* due to the fact that the subject is *this* and, therefore, that the subject is the *first* and *radical* principle of the individuation of accidents.

VIII. – Corollary I: "It is impossible for two accidents that are only numerically distinct to be simultaneously in the same subject."

This is the position of the Thomists against Aureolus, Scotus, Suárez, etc. The Angelic Doctor says: "It is impossible for many accidents of one species to be in the same subject."[151]

The argument is very keen. Accidents of the same species in the same subject would be two and not two at the same time, but this is most manifestly impossible. – They would be two, as it is supposed; but they

[151] *In V Metaph.*, lect. 12 (Text 12): "Impossibile est plura accidentia unius speciei in eodem subiecto esse."

would not be two, because they would be neither formally nor numerically distinct.

– They are not formally distinct, for they are supposed to be of the same species, but those things which coincide in the same species are formally the same. Nor are they numerically distinct, for indeed the principle of individuation and numerical multiplication is the subject. But there is only one subject in this case. Therefore, these two accidents would be numerically one.

– ***Objection***: The same subject can receive many accidents of different species simultaneously. Therefore, it can also receive many accidents that are only numerically distinct. ***Reply***. *I deny the conclusion.* For the subject has diverse potencies in relation to specifically distinct accidents; in fact, it has as many potencies as there are species of accidents in it, but has one potency in relation to one species of accident and to the accidents of one species. This potency cannot be formally diversified except at different times. But the subject taken formally underlies the same action of the same agent at the same time. Therefore, it cannot be formally diversified; therefore, it cannot receive accidents that are only numerically distinct.

IX. – COROLLARY II. Hence we can now understand the very popular axiom: "An accident does not migrate from one subject into another" (*Accidens non migrat de subjecto in subjectum*), that is, it cannot occur that an accident that was in a prior subject passes into another subject while remaining numerically the same, but it is necessary that a new accident be produced. Thus, when motion is propagated from one mobile being to another, there are two numerically diverse motions [that is, one in the mover and one in the moved]. For, since there is a new subject, there is a new and distinct principle of individuation.

THIRD TREATISE

ON THE WORLD
INSOFAR AS IT IS ORDERED TO AN END

In the first treatise, we discussed the world with respect to its efficient cause; in the second treatise, we discussed the world with respect to its material and formal causes; it remains for us to investigate the same world insofar as it is ordered to an end. Now, the world, insofar as it acts and moves towards an end is called 'nature'. The present treatise, therefore, will be on nature. And concerning nature, we can ask *what it is*, *how it is ruled*, and *for what purpose it exists*.

FIRST QUESTION

What Nature Is

We begin with the notion of nature. Nature is the principle of motion and rest; hence the need to discuss motion. Finally, since a thing is known through its opposites, we will add certain observations concerning art and violence, which are opposed to nature.

FIRST ARTICLE

ON THE NOTION OF NATURE[1]

I. – THE MULTIPLE SENSES OF THE WORD. In its primitive sense and according to the etymology of the word, nature is taken to mean the generation of living things, which is called 'nativity' (*nativitas*) or 'sprouting' (*pullulatio*), such that 'nature' (*natura*) sounds like 'about to be born' (*nascitura*). Secondly, this word was derived to signify the very principle of such origins, and, because the principle of living things is intrinsic, it was further derived to signify any principle of motion. Now, this principle is either matter or form: hence, sometimes 'nature' means form and sometimes matter.[2] But the terminus of generation is the composite, or the essence of the species; hence it is that essence and nature are used interchangeably, although in such a way that 'essence' indicates an order to being,[3] because it is a potency towards being, whereas 'nature' indicates an order towards operation. In a third sense, nature is taken as divided into *natura naturans* ('naturing nature') and *natura naturata* ('natured nature'). *Natura naturans* is God himself, insofar as He is the first, universal principle of things, Who rules and governs them and concurs with them in their operations.[4] *Natura naturata* refers to a created substance, insofar as it is the principle of the operations that it elicits of its own accord; or it also could mean the complex of all secondary causes, insofar as they have power to produce determined effects under the ordinary influx of God. Fourth, and especially, nature is taken to mean the principle of acting out of necessity, in opposition to a free cause, which operates at will. Fifth, it is taken to mean one's character, disposition, or temperament; in this

---

[1] One can refer to ARISTOTLE, *Physics* 2.1, 2.5; *Metaphysics* 5.4; ST. THOMAS, his commentaries on these passages in Aristotle; *Summa theologiae* Ia, q. 29, a. 1 ad 4; IIIa, q. 2, a. 1; JOHN OF ST. THOMAS, *Phil. Nat.* q. 9; TOLEDO, SYLV. MAURUS, CONIMBRICENSES, GOUDIN, MAILHAT, and the Scholastics in general, in their *Physics*. J. GREDT, *Elementa*, t. 1, 3rd Ed.

[2] Cf. *Summa theologiae* IIIa, q. 2, art. 1.

[3] Reading *ordinem ad esse*, instead of *ordinem adesse*. – ***The Translator.***

[4] ST. AUGUSTINE, *Contra Julian.*, 1.8, n. 36: "There is absolutely no nature which is not either God Himself or something made by Him" (*Omnino natura nulla est quae non aut ipse Deus sit aut ab ipso facta sit*).

sense, someone is said to be of such a nature as to be inclined to wrath. Sixth, and most specifically, it is taken to refer to the internal, innate, and substantial principle of motion and rest.

II. – The 'Natural'. From the notion of nature we can gather what is *natural*; it is generally defined as: "*That which is according to the requirements, powers, order, and end of nature*" (*Id quod est secundum exigentias, vires, ordinem, et finem naturae*).

– '*According to its requirements*': whatever the concept of the thing requires pertains to its nature and is said to be 'natural'.

– '*And its powers*': all that which a thing can attain through its own powers is connected with nature and can be called 'natural'.

– '*And its order*', for, if something is found in things which is outside of the order that is maintained in the universe, it is miraculous [and not natural].

– '*And its end*', for each thing is specified by its own end; hence, that which is according to its own end, to which nature is driven and to which it tends by its own operation, should be called 'natural'.

Consequently, *supernatural* is that which exceeds the requirements, powers, order, and end of the whole of created nature.[5] Something is said to be *preternatural* if, despite not being above the whole of created nature, it is nonetheless above the mode of some nature.

III. – That which is Opposed to the Natural. The following are opposed to the natural: (1) the *miraculous*, or the *supernatural*, for which reason neither resurrection nor the beatific vision can be said to be natural; (2) the *violent*, hence it is not natural for a stone to be carried upward; (3) the *fortuitous*, or that which is by chance; (4) that which is *free*, for we call a natural agent that which is determined out of necessity to one thing, whereas a free agent is that which is indifferent to many things; (5) the *artificial*, thus the figure of the ox is natural, whereas the form of the house is artificial; (6) the *acquired*, hence our ideas are not natural to us, that is, innate; (7) the *metaphysical*: for we call a natural composite that which consists of matter and form, whereas the metaphysical composite is that which results from essence and existence.

---

[5] On the notion of the *supernatural*, cf. A. Mercier, OP, *Revue Thomiste*, 1902 and 1903.

IV. – The Definition of Nature. In the way that we are using it in the present treatise, the term 'nature' can be defined, according to Aristotle as: "The principle and cause of motion and rest of that in which it is first present *per se* and not *per accidens*" (*Principium et causa motus et quietis ejus in quo est primo per se et non per accidens*.).[6]

It is said together to be (1) 'the principle and cause', for two reasons: first, to signify that in some things nature is a passive principle, whereas in others it is an active principle, which is expressed through the term 'cause'; second, to show that nature is not just any principle, but a first, radical, and positive principle. And to signify this it is not enough to say 'cause', because 'cause' could be also said of a motive power, which is an accident and a proximate principle; nor would it suffice to say 'principle' alone, because this is also applicable to privations. Therefore, so that it may be clear that nature is the first root and origin of motion, and that it truly, positively and physically influences motion, the two words are combined: *principle* and *cause*.

(2) 'Motion' means any physical change that can be apprehended by the senses. We can distinguish three kinds of motion: motion of *alteration* in qualities; motion of *increase* and *decrease* in quantity; and *locomotion*, whereby something is carried from one space to another.

(3) 'Rest'. Here we do not take 'rest' to mean a lack of motion, but as an attainment of the terminus, or as a perseverance in a state that is convenient to the thing, which state is acquired through motion. The meaning is not that nature is always and in all things the same principle of motion and rest, but that in some things it is the principle of motion, and in others the principle of rest, and in yet others of both: for example, in living things it is only the principle of motion, because if motion ceases, life ceases as well; in light and heavy bodies, however, it is the principle of both motion and rest; for it is natural for a body to seek the center of the earth, and for it to rest when it attains it.

(4) 'Of that in which it is present' means that the principle must be internal and innate to the being that moves and rests; hence, these words are added to exclude extrinsic principles of motion, such as art and violence.

(5) 'First' is said to signify that nature is not a secondary and instrumental principle, but a radical and substantial principle.

---

[6] *Physics* 2.1.

(6) '*Per se* and not *per accidens*' is said so that those things which can be from an extrinsic principle, but are joined to the thing in question *per accidens*; as in the case of a physician who heals himself, where the act of curing comes from an extrinsic principle that is nonetheless conjoined *per accidens*, for the one who is healed happens to be the physician; nor does the physician heal himself as physician, but as a patient; hence, such an act of healing is not said to be by nature, but by art. This is the unanimous view of the Scholastics.

V. – Things To Which the *Ratio* of a Nature Applies. It is first of all form which has the *ratio* of a nature. For form is the principle of motion, for example, the motion of nutrition, or of growth, in living things, or the motion of attraction, etc., in minerals. It must be noted that form is the principle of those motions that perfect the subject, but not the principle of corruption, or of the corruptive alterations of the subject in which it is; for nothing tends to its own corruption.

Matter also has the *ratio* of a nature, because through its transmutation it is truly a principle of motion and of rest, and it is a first and radical *ratio* that renders the substantial whole apt to be moved with the motions of generation, alteration, etc.

The composite whole taken as a supposit or person does not have the *ratio* of a nature, because thus it is more a *principiatum* (lit., 'a principled [thing]') than a principle (*principium*), and is really distinct from nature, as we shall say in *Metaphysics*. But the composite taken as an essence, or as a species, can to some extent be said to be a nature, because it is a substantial and radical principle of its motion; but nonetheless, since it presupposes other natures that are prior to it, namely, matter and form, it is not *simpliciter* a nature.

The rational soul, in its vegetative and sensitive degrees, is a nature, just like other sensible forms, and is the principle of the motions that are apprehended by the senses. But what should we say concerning the soul as rational? It is also a nature in this way: for the constitutive and specifying form of man is the intellectual soul itself. But man is a natural *ens*. Therefore, the soul, as intellectual, is a nature. In fact, the separated soul retains the *ratio* of nature due to its connaturality to operate in a body and according to the dispositions of the senses.

Angels and God have the *ratio* of a nature, if 'nature' is taken in the general sense of a quiddity or essence; but in no way are they natures in the specific sense in which we take it in the present treatise. For the term 'nature', as it is used in Physics, is the principle of sensible motion, or of divisible and imperfect motion. But such a motion cannot be ascribed to God. Angels also elicit perfect operations, and are not subject to physical and sensible motion.[7]

VI. – SOME OF THE PROPERTIES OF NATURE. The sayings of the ancients concerning universal nature are well known; here we give several of them.

– "Nature does not make leaps, but ascends from the more imperfect to superiors, in such a way that the highest part of the inferior attains to the lowest part of the superior" (*Natura non facit saltus, sed ab imperfectioribus ascendit ad superiora, ea quidem ratione ut supremum inferioris attingat infimum superioris*). Thus, the highest of the plants attains to the level of the lowest of the animals; and the highest of the senses, namely, the estimative power, attains somewhat to the lowest level of the intellect; for we see that in the more perfect animals there is a certain imitation of judgment and reasoning.

– "Nature does not do anything in vain, nor does it do through many things what can come to be through few things. Nature is a lover of unity" (*Natura nihil facit frustra, nec facit per plura quae per pauciora fieri possunt. Natura est amans unitatis*). It reduces many actions to one potency, and many potencies to one essence.

– "Nature makes those things which are divided in inferiors be united in superiors." Thus, those things which are divided in minerals are more united in plants; for in minerals the terminus of operation is outside of the operating principle; but in plants the terminus remains in the agent itself. There is still a higher unity in sensitive things than in vegetables; for indeed, the terminus of a vegetative operation remains in the same subject, but not in the same faculty: for example, the terminus of nutrition is in the whole subject, and not in the nutritive faculty; but on the other hand, the terminus of a sensitive operation is in the same subject and in the same faculty: for example, the terminus of vision is in the visual potency. Unity in more perfect in an internal sense than in an external sense, for the common sense attains of itself the objects of the five external senses. Unity is more perfect in the intellect than in the senses, for the intellect of itself and by a single idea knows that which the external senses and the internal senses, taken together, apprehend through many species. Higher still, namely, in the angel, there is a yet

[7] Cf. JOHN OF ST. THOMAS, I. P. *Phil. Nat.*, q. 9.

greater unity: for the angel understands through a single idea many things that are represented in us through many species. And the more superior an angel is, the fewer and more universal are the ideas that it receives. Finally, in God, the author of nature, there is the highest unity: for existence, essence, faculty, operation, idea, are one and the same thing.

– "Nature neither lacks in necessary things nor exceeds in superfluous things; but yet acts for abundance, and not for poverty" (*Natura non deficit in necessariis, nec excedit in superfluis; attamen agit ad opulentiam non ad paupertatem*).

– "Nature aims at the more perfect; hence, the more perfect some things are, in so much greater an excess are they created by God" (*Natura intendit perfectiora; unde quanto aliqua sunt magis perfecta, tanto in majori excessu sunt creata a Deo*).[8] For indeed, nature must intend and attain that which is the end of creation. But the end of creation is the perfection of the universe. Therefore, it is necessary that the more perfect things in nature be produced in greater excess. In bodies, this excess consists in magnitude; hence, the more perfect bodies exceed others almost incomparably according to magnitude; whereas in spiritual substances, in which there is no magnitude, excess is considered according to multitude. Hence, it is necessary that immaterial substances exceed material substances almost incomparably according to multitude.

The aforesaid dictum is understood of the principal parts of creation, or of the degrees of *ens*, which are *esse* in bodies, living in plants, sensing in animals, reasoning in man, and understanding in angels; and of beings that are aimed at for their own sake, namely, species as well as individuals are equivalent to own species by reason of their nobility. But nature does not aim at corruptible individuals *per se*, but only for the sake of the conservation of the species.

Hence, it is fitting that plant species exceed the species of bodies; and that animal species exceed plants species. But since man is in a higher degree, and because each human individual is aimed at *per se* and is equivalent to one species, it is necessary that the number of men at some point exceed the number of animal species. Finally, the number of angels must exceed the number of men almost incomparably.

The aforesaid axiom is confirmed by experience and by science, for there are about 70 species of simple bodies, and the number of

[8] Cf. *Summa theologiae* Ia, q. 50, a. 3.

composite bodies [i.e., chemical compounds] is about 300, whereas there are 12,000 species of plants, or according to some scientists, about 20,000; there are 400,000 species of plants, or in fact, as many estimate, about 700,000. And the number of men incomparably exceeds that of the other species.[9]

– "Nature always makes the best as far as possible" (*Natura semper facit melius quoad potest*). "It must be said that nature does not always make what is best to have with respect to each part, but what is best to have with respect to the whole; otherwise, it would make the whole body of man be an eye or a heart; for this would be best for each part, but not for the whole. Similarly, even though it would be best for something to be placed in a higher order, nonetheless it would not be best for the universe, which would remain imperfect, if all creatures belonged to the same order."[10]

----

[9] By 'simple bodies' Hugon is here referring to the number of chemical elements in the periodic table. Hugon is basing his estimate of the number of elements on the original table of elements which, at the time of writing (1905), had just been published by the chemist Dmitri Mendeleev, and which contained 63 elements. As of 2012, the current standard table of elements, has 118 elements, 98 of which occur naturally. By 'composite bodies' Hugon means chemical compounds; in Hugon's time, only some 300 chemical compounds were known, whereas today we know of 200,000 real chemical compounds existing in nature (about 60,000,000 total compounds are known but this number includes synthetic compounds made in laboratories, most of which do not exist in nature). Current scientific estimates for the number of 'plant' species in existence is 1 million (this number includes non-sentient organisms, which is the meaning of 'plant' in Aristotelian-Thomism; within the taxonomic group *plantae* there are only 300,000). The current estimated number of animal species is 7 million. The number of humans living on Earth today is estimated to be about 7 billion. Thus, the principle still obtains today. And we must remember that this is a principle of fittingness; therefore, it would be unfitting, but not impossible, for the principle not to obtain in reality. – ***The Translator.***

[10] ST. THOMAS, *De potentia* q. 3, a. 6, ad 26: "[D]icendum, quod natura non facit semper quod melius est habito respectu ad partem, sed habito respectu ad totum; alias totum corpus hominis faceret oculum vel cor; hoc enim unicuique partium melius esset, sed non toti. Similiter licet melius esset alicui rei quod in altiori ordine poneretur, non tamen esset melius universo, quod imperfectum remaneret, si omnes creaturae unius ordinis essent."

SECOND ARTICLE

ON MOTION[11]

I. Definition of Motion. Since nature is the principle of physical motion, the consideration of motion logically follows the discussion on the notion of 'nature'. Now, the Aristotelian definition of motion is well known: "The act of something that exists in potency as such, that is, insofar as it is in potency" (*Actus existentis in potentia, in quantum hujusmodi, seu prout in potentia*). In order to understand this, we must recall that something can exist in any one of three states: in potency only without being reduced to act; in its perfect and last act; or midway [between potency and act], that is, partly in act and partly in potency. When a thing is in potency only, it is not said to be moved; whereas when it is in perfect act, it is no longer moved, but was moved [in the past]. Therefore, for it to be said to be moved, it must be somewhat in act and nonetheless to remain in potency to a further act. For this reason, motion should not be defined as, "the potency of something that exists in potency," or, "the act of something that exists in act," but rather, "the act of something that exists in potency." It is said to be an *act*, in the manner of the genus of the definition, and here act is taken according to the common meaning insofar as it is opposed to potency and abstracts from the perfect and the imperfect. But motion has a twofold relation: one to the subject in which it is, and the other to the terminus to which it tends. The first relation is indicated by the words, "of a being in potency," namely, of a subject that exists in potency; the second relation is indicated by the words, "insofar as it is in potency," that is, insofar as it actually tends to a further act. But if this actual tendency ceased, the act, although imperfect, can no longer be called motion. For example, water begins to be heated at ten degrees; as long as it tends to the terminus, it moves; now, supposing that the heating stops at five degrees, there will be an act of a subject that exists in potency to a further degree of heat, but it will not be motion, because the actual tendency to the terminus has stopped. Hence, it is not

[11] One can consult Aristotle, *Physics* 3.1; St. Thomas, *In III Phys.* lect. 5; John of St. Thomas, Sylvester Maurus, and other Scholastics already cited; Secchi, *L'unité des forces*; Farges, *Le moteur et le mobile*; Nys, *La notion de l'espace*; Duhem, *Évolution de la mécanique*.

sufficient to define motion as, "the act of something that exists in potency," but it must be added, "insofar as it is in potency," that is, insofar as it is actually ordered and tends to a further act.

– There is another definition given by the Philosopher, which coincides with the preceding definition: "Motion is the act of a mobile thing insofar as it is mobile" (*Motus est actus mobilis in quantum est mobile*). The Angelic Doctor explains the reason of the definition: "Motion is the act of something that exists in potency insofar as it is such [i.e., insofar as it is in potency], but that which exists in potency insofar as it is such is mobile, and not a mover, because a mover insofar as it is such is an *ens* in act; it follows that motion is the act of something mobile insofar as it is such."[12]

II. – What Belongs to the *Ratio* of Motion. *Continuity* pertains to motion. But discrete motion is nothing other than many continuous motions that are interrupted; so it is not a new species of motion, but is referred to the species of continuous motion. *A transition into something external* belongs to the *ratio* of motion; for motion is conceived as an actual tendency to something ulterior. Hence, immanent actions, which remain in themselves and have themselves as terminus, are not properly motions.

Terminus pertains to motion as something that specifies it, but it does not intrinsically constitute the very essence of motion, since motion can remain motion without the mobile thing reaching its terminus.

III. – How Motion is Related to Action, to Passion, and to the Terminus. In motion there are four relations: to action, to passion, to the *terminus a quo* ('terminus from which'), and to the *terminus ad quem* ('terminus to which'). But is motion distinct from these things? If so, how? Many authors, especially among the Scotists, think that motion is really distinct from action and passion; others, such as Suárez, admit only a distinction of reason. The Thomists generally hold that between action and passion there is neither merely a distinction of reason, nor a real-entitative distinction, but a modal distinction.

---

[12] *In III Phys.*, lect. 4: "Quia enim motus est actus existentis in potentia inquantum huiusmodi; existens autem in potentia inquantum huiusmodi, est mobile, non autem movens, quia movens inquantum huiusmodi est ens in actu; sequitur quod motus sit actus mobilis inquantum huiusmodi."

Proof of this View. A distinction of reason does not suffice when things differ before the mind conceives them. But motion is distinct from action and passion independently of the operation of the mind. *Therefore. Proof of the Minor.* An *ens* that is really incomplete before the operation of the mind differs from one which is really complete. But action and passion are complete beings, each of which is directly placed in a predicament [i.e., in one of the ten categories]; but motion is a being that is really incomplete, namely, something that flows and journeys, which cannot be directly referred to a predicament. *Therefore.*

Further, the coming out of an accident from the agent, the reception of the effect in the patient, and the actual tendency towards the terminus are not mere concepts, but relations that are at least modally distinct. But an action is the coming out of an effect from an agent, passion is the reception of an effect in the patient, and motion is an actual tendency towards a terminus. Therefore, these three things are at least modally distinct.

– Nonetheless, there is not a real-entitative distinction. For an entitative distinction involves the addition of a new reality. And motion does not add a new reality, but is only the modification of the same reality, which is an action insofar as it comes out of the agent, which is a passion insofar as it is received in the patient, and which is called 'motion' insofar as it is an actual tendency towards the terminus.

The same is true of the distinction between motion and the terminus. First, motion is distinct from the *terminus a quo*. For indeed, those things of which one ceases to be when the other begins to be are not the same. But when motion begins, the *terminus a quo* ceases to be, or is left behind. *Therefore.*

Second, it is distinct from the *terminus ad quem*. For motion is a fluid and transitional *ens*, but the terminus is something absolute; thus heat, which is the terminus of being heated, is in itself something absolute. Therefore, motion is at least modally distinct from the *terminus ad quem*. – Further, a form receives a different modality when it is being acquired and when it is possessed. But motion is a form that is being acquired, whereas the *terminus ad quem* is a form that is [already] acquired and possessed. Therefore, motion and the terminus are different modalities.

IV. – THE PREDICAMENT TO WHICH MOTION BELONGS. Motion, insofar as it is an incomplete *ens*, does not constitute a special predicament; but if it is considered as something that denominates a moving thing or a moved thing, an agent or a patient, it pertains reductively to the predicaments of action and passion. If it is taken as an act that is related to its terminus as something imperfect to something perfect, it is referred reductively to the predicament of its terminus: for example, motion to heat is reduced to the predicament of quality; for indeed, the imperfect and incomplete is reduced to the predicament of the perfect and complete. Finally, if it is seen as a certain state that is opposed to rest, it is not reduced to a predicament, but is a post-predicament, namely, a *ratio* that is found in many predicaments.

V. – The Subject of Motion. The common opinion (*sententia communis*) is that the mobile thing is the subject of motion.

Proof. Motion is of the same entity as the terminus. But the terminus is received in the mobile thing; for example, water is clearly subject of the terminus of heating. *Therefore.*

Further, an accident is received in that of which it is an accident or property. But motion is an accident or property of a mobile thing. Therefore, its subject is the mobile thing.

Thus, the terminus of alteration, which is quality, is received in the thing that is altered, and the terminus of increase in that which is increased. The Scholastics agree on this, but there arises now the celebrated controversy concerning the subject of action, which is modally distinct from motion.

VI. – On the Subject of Action. The difficulty concerns transient action alone, for immanent action does not belong to motion, nor is it reduced to the predicament of action, but rather to quality. Moreover, all acknowledge that the subject of an immanent action is the agent itself. But philosophers, even the Thomists among themselves, dispute concerning transient actions. The Scotists claim that the subject of the action is the agent itself. For they argue thus: The relation of the agent must be in the agent and the relation of the patient in the patient. But an action is a relation of the agent and a passion the relation of a patient. Therefore, the subject of an action is the agent, and the subject of a passion is the patient. Cajetan, Nazarius, and many other Thomists hold this view, not on account of Scotus' argument, but on account of another motive, which can be proposed thus: The Agent intrinsically perfects and changes that which is in the agent itself as in its subject.

But an action is a second act by which the agent intrinsically changes and perfects. Therefore, its subject is the agent. Another, contrary view holds that the action is subjectively in the patient. This is the view of Capreolus, Hervaeus Natalis, Ferrara, Soto, the Complutenses, Suárez, and Mailhat. Others, finally, in order to reconcile these views, make a distinction between two formalities in an action: one whereby the agent is actuated and reduced from habit to act, and another which is in the terminus and reduced the patient to act; hence they conclude that action is inchoative and originative in the agent, but consummative in the patient or in the effect. This is the view of John of St. Thomas, which Goudin accepts. We have chosen this latter view. Hence it is that:

VII. – CONCLUSION: "An action, taken causally, is in the agent; but taken strictly and properly, it is subjectively in the patient."

– ***Proof of the First Part.*** An action taken causally is an *actuality of power*[13] and a *complement of a potency.*[14] But it is clear that the actuality of a power is in the power itself that it actuates and that the complement of a potency is in the potency itself that it completes. Therefore, an action taken in this way is in a power or in an operative potency, that is, in the agent itself.

Scotus' argument would be good if the action were a relation between the agent and the patient, but an action is not a relation; rather, relation follows action and is founded in action.

– ***Proof of the Second Part.*** An action taken properly is actuality insofar as it brings about an effect. But an actuality that brings about an effect is in the effect itself. Therefore, an action taken properly is in the effect, or in the patient. *Proof of the Major.* An action taken properly is a formality whereby something is properly denominated as an agent. But something is considered to be properly an agent, not on account of the action that is in the cause, but insofar as it brings in an effect in the terminus; thus, fire is said to heat, not from the actuality that is in the fire, but from the actuality according to which it brings heat in the water. Therefore, an action taken properly is an actuality that brings in an effect. *The minor is clear.* An actuality that brings in an effect is a diffusion of an act into the patient. But the diffusion of an act into a patient is, of course, in the patient. Therefore, an actuality that brings an effect is received in the patient.

---

[13] Cf. *Summa theologiae* Ia, q. 54, a. 1.

[14] Cf. *Summa contra gentiles* 2.9.

– ***Confirmation***. An action is of the same entity as motion. But motion is received in a mobile thing or in a patient. The Angelic Doctor discusses the issue: "Even though the action is from the agent as that from which it comes (*a quo est*), it is nonetheless in the patient as received in it."[15] Through this argument, the diverse opinions are reconciled and the difficulties that are usually brought up are resolved.

VIII. – WHAT CAN BE THE TERMINUS OF MOTION. We have noted many times that successive motion, which we are now discussing, is a transitional tendency toward a terminus. Now we inquire what can be such a terminus. It must be noted first of all that three conditions are required in the terminus of motion. The first is that a terminus have some breadth along which [the mobile thing] is to advance. For indeed, motion is not a sudden acquisition of a terminus, but one that is successive and gradual, and which occurs in different parts of time. But it can only occur gradually and in different parts of time if between the two termini there is some intermediary breadth: either (a) an *extensive* breadth, such as quantity, locus, or place, by reason of which the terminus is slowly acquired, one part at a time—such is the breadth that exists in the motion whereby someone travels from Italy into France; or (b) an *intensive* breadth, such as quality, by reason of which the terminus is acquired through degrees or through greater or lesser modes of being rooted in the subject—such is the breadth that exists in motion whereby water transitions to one hundred degrees of heat.

The second condition follows from this, namely, that the *terminus a quo* and the *terminus ad quem* be positive: for privation, since it is indivisible, is not acquired or left behind in parts and step by step, but altogether simultaneously and in an instant. And finally, the third condition follows, namely, that the two termini be opposed with contrary opposition. For contrary opposition is that between two positive termini that are absolute and cannot possibly exist together. But the termini of motion are positive, as was said; they cannot possibly exist together, for when the *terminus ad quem* comes, the *terminus a quo* necessarily ceases to be; and they are absolute, not relative, because the relation does not occur *per se*, but results for the sake of the production of its terminus and foundation. Therefore, the termini of motion are opposed with contrary opposition. Having fully understood these conditions, it

[15] Cf. *In III Phys.*, lect. 5.

will become apparent that motion cannot have many predicaments as its subject. Hence it is that:

IX. – CONCLUSION: "*Substance*, *relation*, *action*, *passion*, *'when'*, and *habit*, cannot be the terminus of motion; rather, motion can only occur *per* se with respect to *quantity*, *quality*, and *where*."

(1) *With respect to substance.* Motion is between two positive and contrary termini. But substance does not come to be from a positive and contrary terminus, but from its privation. Therefore, substance is not the terminus of motion. Further, motion is the successive acquisition of a terminus in parts. But since substance is indivisible, it does not come to be in parts or successively, but altogether simultaneously and in an instant; hence the axiom: "The generation of a substance occurs in an instant" (*Generatio substantiae fit in instanti*). Therefore, there is no motion *per se* towards substance. We say '*per se*' because substance can be the terminus of substance, insofar as it is affected by quantity, or has some alteration joined to it: for although generation occurs in an instant, nonetheless the dispositions and alterations that precede it occur with motion and successively.

(2) *Action and passion are not the terminus of motion.* For they are the same reality as motion; hence, if there were motion toward an action, there would be motion towards motion, or action toward action, and thus there would be an infinite regress.

(3) *Relation, position, 'when', habit, cannot be the terminus of motion.* They do not have contrariety or breadth, whether intensive or extensive; they do not come to be successively and by reason of themselves, but result in an instant by reason of something else: *relation* for the sake of the production of the terminus and of the foundation; *position* from *where*, or from the disposition of the parts in a place; *'when'*, due to the fact that the thing is in time as in an extrinsic measure of its own duration; *habit* from the extrinsic application of vestments or armor.

(4) *Quantity, quality, and 'where' can be the terminus of motion.* We acknowledge that these three are not always the terminus of motion, but sometimes come to be in an instant, as when quantity or quality are produced in the generation of a substance; but the sense of the conclusion is that these things, *per se* and according to their connatural mode, can be the terminus of motion. For indeed, they have the three required conditions. First, they receive breadth, along which they

successively advance: and this breadth can be extensive breadth, as that between a great and a small quantity; or intensive breadth, as that between the diverse degrees of quality, for example, that between the one and one hundred degrees of heat; or the breadth of local distance, as that between Lyons and Rome, or between a place above and a place below. Second, the two termini are positive: a greater and a lesser quantity, one and one hundred degrees of heat, and a place above and a place below, are all positive things. Third, the termini are incompossibile and contrary, as is clear in the case of greater and lesser quantities, of more intense and less intense qualities, and of the higher and the lower.

X. – WHENCE MOTION DERIVES ITS UNITY AND DISTINCTION. Motion does not have a proper and rigorous unity and distinction, because it is an incomplete being; the question, then, has to do with an improper and reductive unity, which could be generic, specific, or numeric.

XI. – FIRST CONCLUSION: "Motion derives its generic and specific unity from its *terminus ad quem*."

Proof. A thing that is essentially ordered to another is diversified and specified by that to which it is first *per se* related. But motion is something that is essentially ordered to something else; it is first *per se* related to the *terminus ad quem*. Therefore, motion is diversified and specified by its *terminus ad quem*. *The major is clear*; for, when the whole *esse* of a thing is for the sake of something else, it must be measured in relation to it. *Proof of the Minor*. Motion is essentially a tendency, a way, or an approaching of the *terminus ad quem*. Therefore, it is essentially ordered to it, and it is first *per se* related to it; whereas the *terminus a quo* relates to it only by reason of the *terminus ad quem*; for the *terminus a quo* is left behind only for the sake of approaching the *terminus ad quem*.

But since there are three genera, or predicaments, which are the termini of motion, it follows that there are three genera of motions; the first is motion toward quantity, and is divided into *increase* and *decrease*; the second genus is motion to quality, and is divided into *alteration*, *intensification*, and *remission*; the third is motion to 'where', and is called locomotion, which is subdivided into *rectilinear*, *circular*, and *mixed.*

XII. – Second Conclusion: "For the numeric unity of motion, the following three are necessary and sufficient: the numeric unity of the mobile thing, the numeric unity of the *terminus ad quem*, and the uninterrupted unity or continuity of time."

Proof of the First Part. Motion is an accident. But an accident takes its numeric unity from the subject. Therefore, the unity of the subject is required for the numeric unity of motion. Now, the subject of motion is the mobile thing. Therefore, for the numeric unity of motion, the numeric unity of the mobile thing is required.

Proof of the Second Part. Motion is an essential tendency toward the terminus, such that it is entitatively identical with the terminus itself. Therefore, the unity or plurality of the terminus produces a unity or plurality in the motion.

Proof of the Third Part. Motion is something continuous. But for the unity of a continuum it is required that there be no interruption present. Therefore, the unity of motion requires that there be no rest or interruption of time present within it.

XIII. – On the Contrariety of Motion. We shall cover the issue of the contrariety of motion by making a few points.

(1) The contrariety of motion arises formally from the contrariety of the *terminus ad quem*. ***Proof.*** The opposition of things follows upon their species as a property. Therefore, motions receive their opposition or contrariety from that from which they take their species. But motions derive their species from their *terminus ad quem*. Therefore, they also derive their contrariety from their *terminus ad quem*.

(2) The same mobile thing, according to the same part, cannot be moved *per se* with two contrary motions. ***Proof.*** Two motions are contrary to each other insofar as one is an approaching towards something and the other is a receding away from that thing; for example, heating is an approaching towards heat, and cooling is a receding away from heat. But it is impossible for the same body, according to the same part, to approach towards and recede away from the same terminus simultaneously and *per se*. Therefore, it is impossible for the same mobile thing, according to the same part, to be moved *per se* with two contrary motions.

(3) The same mobile thing can simultaneously be moved with contrary motions according to different parts; or even according to the whole mobile thing, if one motion belongs *per se* while the other only *per accidens*, or both *per accidens*. ***Proof*.** Wherever there are different parts there are different subjects. Therefore, wherever there are different parts, there can be contrary motions. Thus, one hand can be moved downward and the other upward; the foot can move with a motion of cooling while the hand can move with the motion of heating.

The second part of this third point is also clear. We see that a contained body is moved according to the motion of the container, and thus it can possess two motions: one motion that is proper and *per se*, and another that is *per accidens* and which corresponds to the motion of the container. In fact, it often happens that, when the container is moved according to one part, the contained object is moved according to the opposite part: for example, when a boatman rows harder on the east, the boat is moved westwards. Therefore, then the same mobile thing has two contrary motions: a proper motion toward the east and another, *per accidens* motion toward the west. This is the unanimous view of the Scholastics.

(4) The same mobile thing can be moved simultaneously with a motion of rotation about itself, and with a motion of translation about its center, as the Earth is moved about itself and about the Sun. These motions do not imply a contradiction, but rather are ordered in a wonderful way. For while the Earth is moved about itself, it changes its place, and by successively leaving its place it can approach nearer the sun or recede away from it.

XIV. – ON THE THEORY OF MOTION. It is not our job to consider the scientific theory of motion; moreover, the experts do not have a single explanation with which they all agree. But the philosophical theory is contained in the following assertions, which are almost axioms:

(1) *Motion requires a mover, a mobile thing, time, and two termini.* The Angelic Doctor explains the matter:

> "Motion requires five things. First, a first mover is required from which the beginning of motion arises for everything that moves is moved by another. Second, a mobile thing is required, which is what is moved. Third, time is required because motion occurs successively, and thus in time. And beyond these three, two termini are required: namely, one

from which motion begins, and another towards which motion proceeds: for all motion is from something towards something."[16]

Between two termini there must be a certain breadth along which [the mobile thing] successively advances, as we explained in n. VIII.

(2) *Motion is properly the act of a mobile thing.* For it is the act of something that exists in potency; and that which exists in potency is the mobile thing itself.

(3) *The mobile thing can be merely passive or also active.* Indeed, every mobile thing, insofar as it is mobile, is passive, for nothing is moved except insofar as it is in potency or in passivity, although the mobile thing can in another respect be active and react. Therefore, a mobile thing that is merely passive receives only an impulse from the mover according to its mode of capacity, nor is it in any way capable of changing the influx of the agent. But an active mobile thing, at the same time that it undergoes the change, is excited to act, and exerts its own energy and reacts in the mode of its nature, as a guitar that is moved by a guitarist reacts and emits sounds.

(4) *A mobile thing at rest cannot give itself motion, and neither can a mobile thing in motion modify its motion.* This is the well-known law of inertia,[17] which the unanimous judgment of scientists has approved and which serves as the basis for all of physical science. Its reason is clear. To transition from rest to motion, or even to modify motion, is to give oneself an act or a mode of an act. But the mobile thing, insofar as it is an *ens* in potency, cannot give itself an act that it in no way has. *Therefore.*

(5) *Rest is opposed to motion, but only that rest which is in the terminus a quo.* Rest is the privation of motion. Therefore, it is privatively opposed to motion. Not all rest, however, is opposed to motion, but only that which is in the *terminus a quo.* For indeed, motion is the cause of the rest that is in the *terminus ad quem.* But something cannot be the cause of its opposite. Therefore, the rest that is opposed to

---

[16] *In V Phys.*, lect. 1: "[A]d motum requiruntur quinque. Primo requiritur primum movens, a quo scilicet est principium motus. Secundo requiritur mobile quod movetur. Tertio, tempus in quo est motus. Et praeter ista tria requiruntur duo termini; unus scilicet ex quo incipit motus; et alius in quem motus procedit: omnis enim motus est a quodam in quiddam."

[17] Concerning inertia, see NYS, *Cosmology*, n. 205.

motion is not the rest of the *terminus ad quem*, but only that of the *terminus a quo*.

(6) *The motion that is propagated from one thing to another is not numerically the same.* For, since an accident is individuated by its subject,[18] the accident that is received in the new subject cannot be numerically the same [as the one in the previous subject]. But the propagated motion is an accident that is received in a new subject. Therefore, it is not numerically the same [accident].

– Moreover, scientific experiments establish that the motion of the mover is partly lost during its transmission; therefore, it cannot be identified with the new motion that is propagated.

(7) *The motion of the mover precedes the motion of the mobile thing in* ratio *and causality.* This is clear, for the motion of the mover causes the motion of the mobile thing; it is, therefore, prior to it, as a cause is prior to its effect.

(8) *Locomotion is the most perfect of all motions*; for in locomotion the mobile thing undergoes a less significant change, because it is not in potency to something intrinsic, insofar as it is such [i.e., insofar as it is intrinsic], but only to something extrinsic, namely, a place.

(9) *Motion consists of continuous and divisible parts, and at the same time requires both continuative and terminative indivisible parts.* This proposition is clear from what we said above concerning the continuum.[19]

(10) *Motion is infinitely divisible.* This was established when we discussed the divisibility of the continuum and of quantity.[20]

(11) *Motion begins and ceases through an extrinsic instant.* Recall what we said concerning the duration of successive beings.[21]

(12) *Everything that is moved is moved by another.* This dictum is explained in *Ontology*, in the Treatise on Causes, q. II, a. II.

There are other respects in which motion is logically connected with the notions of potency and act; these are more appropriately discussed in *Ontology*, Treatise I, in the question On Potency and Act.

[18] Q. 3, a. 2.
[19] *Ibid.*
[20] *Ibid.*
[21] Q. 3, a. 5, n. 5.

THIRD ARTICLE

ON ART AND VIOLENCE[22]

I. – WHETHER ART CAN BRING ABOUT THE WORKS OF NATURE. The term 'art' has two meanings: (1) as the artificial form that an extrinsic agent induces into some subject; (2) as the habit of art, which is defined as "right reason concerning things that can be made" (*recta ratio facibilium*).[23] Hence, the meaning of this question is twofold: (1) whether an artificial form can bring about the works that proceed from a natural form; and (2) whether an intellectual agent, in virtue of his art, can bring about the works that are produced by natural agents.

II. – FIRST CONCLUSION: "Art, understood as artificial form, cannot in any way carry out natural works."

Proof. An intrinsic principle of motion is opposed to an extrinsic principle of motion. But nature is an intrinsic principle of motion, whereas an artificial form is an extrinsic principle. Therefore, nature and artificial form are opposed, and thus it is impossible for the form of art to be able to bring about the very operations that nature elicits. *The minor* expresses the notions of nature and art, and is self-evident. An artificial form is impressed by the guidance of the artificer. But the artificer is extrinsic. Therefore, the form, insofar as it comes from art, receives only the *ratio* of 'extrinsic principle'.

III. – SECOND CONCLUSION: "Art, taken as the habit of art, or in other words, as a rational artificer who acts in virtue of his art, cannot immediately by his own power bring about the works of nature; but it certainly can bring them about mediately and through the power of nature, by applying active powers to passive powers."

Proof of the First Part. In order for the artificer to be able to do the works of nature immediately and by his own power (*propria virtute*), he must contain them virtually (*virtute*). But a created artificer neither

[22] Cf. ST. THOMAS, *De occultis operibus naturae* (*Opusc.* 34); JOHN OF ST. THOMAS, I. P. *Phil. Nat.* q. 9; GOUDIN, and in general the Scholastics in their *Physics*.

[23] See what was said concerning art in *Minor Logic*, section "Preparation for Logic."

includes nor carries in his power (*in sua virtute*) the works of nature, namely, the substantial forms that nature produces, as is clear that in men fire is not contained virtually. Therefore, a created artificer cannot immediately and by his own power bring about the works of nature. ***Confirmation.*** To do the works of nature immediately, for example, as substantial forms do, is to have within oneself a formal principle for bringing about those forms or at least to be able to supply such a formal principle. But a created artificer neither has in himself a formal principle for bringing about forms nor can he supply that formal principle; for indeed, a created artificer does not change bodies formally, nor do the substantial forms of things flow from it. *Therefore.*

Proof of the Second Part. Through the direction of art, a rational agent can apply active natural powers to passive natural powers. But a natural work follows the application of a natural agent to its corresponding passive power. Therefore, a rational artificer can in this way bring about the works of nature.

Thus a physician can restore health, which is a work of nature, by applying active powers; thus the angels, even though they cannot give life by their own power, can nonetheless apply seeds and, by using the power of such seeds, produce living things.

IV. – EXPLANATION OF THE DICTUM: "ART IMITATES NATURE" (*ARS IMITATUR NATURAM*). A created artificer must, as far as it is able, assimilate itself to the first agent. But this assimilation occurs through the imitation of nature: for nature is the artifact [or art-product] of the divine Wisdom, and is like a book in which we read the divine ideas and from which we derive our own ideas. Therefore, art must imitate nature.

This imitation consists in four things: (1) The work of art, just like the work of nature, is a work of intelligence; for just as the divine Wisdom is the right reason of those things which are done through nature, so the intellectual habit that resides in the mind of the artificer is the right reason of those things that are done through art. (2) Art, like nature, acts out of something that is presupposed: that is, nature does not draw out anything out of nothingness, but only by changing preexistent matter; thus, in its work, art presupposes matter, which it changes, and by changing it, it raises it to something superior. (3) Art, like nature, proceeds in an orderly way, that is to say, it adapts adequate means to the end intended. (4) Art, like nature, must act for an end and tend to an end; hence, just as a monster is produced when nature falls away from its end, so an error is produced when art deviates from its end.

V. – How Nature Surpasses Art. Nature surpasses art in the same four respects: (1) Nature comes from the most universal and most efficacious Reason, whereas art comes from a particular and fallible reason. (2) Nature, by changing preexisting matter, can immediately produce a substantial form, whereas art can do so only mediately by using the power of nature. (3) Nature proceeds in a more certain and less fallible manner, and employs more efficacious and more congruous means. (4) The end that nature intends is the universal good of the whole universe; whereas the end of art is a particular and limited good. This view is the general view of the Scholastics.

VI. – Violence. The violent is commonly defined, according to Aristotle, as: "That which comes about from an extrinsic principle, the patient not contributing its power" (*Quod est a principio extrinseco, passo non conferente vim*).[24] Two things, therefore, are required in the concept of violent [motion]: (1) that the principle of such a motion be neither the essence nor the principles of the essence [of the patient], but some external agent; (2) that the patient contribute nothing, that is, that it does not supply an inclination towards that motion. Now, a patient can 'contribute nothing' either negatively or positively: *negatively*, if it is neither inclined nor resists, but is related indifferently [to the motion]; *positively*, if it has a contrary inclination, and if, as far as it can, it resists an extrinsic principle. For motion takes away only the negation if it is in a subject that is related negatively [to the motion]. And if the negation is taken away, there is nothing that prevents the placing of the form. Therefore, extrinsic motion is not violent motion if it is in a subject that is related negatively to the motion. Therefore, the meaning of the words "the patient not contributing its power" is: "the patient withstanding and resisting." Again, it must be noted that the patient sometimes, indeed commonly, resists actively, when it has in itself an active principle that is opposed to the violent principle, and then it is said to undergo *active violence*; sometimes, however, it resists passively, when it is moved from its natural state, and is prevented from receiving its proper forms, or from obeying its connatural active principle; and then it is said to undergo *formal violence*.

[24] ARISTOTLE, *Ethics* 3.1. Cf. ST. THOMAS, *In VIII Metaphys.*, lect. 7.

VII. – Whether God can Introduce Violence in His Creatures. There are three opinions on this matter. Certain thinkers hold that God, being the most universal agent, can introduce violence; others, on the other hand, claim that God can never introduce violence, even if he wishes to operate in the mode of a particular agent. An intermediary opinion, which many Thomists follow and which we have chosen, can be proposed thus:

VIII. – Conclusion: "God, insofar as He is the universal mover, cannot introduce violence; but He certainly can, if He intends to operate as a particular mover."

Proof of the First Part. That which is according to a creature's stronger inclination is not violent to it. But whatever God as universal agent can introduce in a creature is according to that creature's stronger inclination. Therefore, it is not violent to that creature. *Proof of the Minor.* Any creature has a twofold natural inclination: one towards its own good, and the other to obey the Creator in all that He wills to do with His creature. And, just as the part is more inclined toward the good of the whole than toward its own good, thus the inclination of the creature to obey the most universal Agent is stronger than its inclination toward its own good.

Proof of the Second Part. In order for God, insofar as He is a particular mover, to introduce violence onto his creatures, it is sufficient that He preserve the natural inclination of some creature while, nonetheless, introducing something contrary to this inclination. But it is not impossible for God to conserve the natural inclination of a creature while doing something contrary to this inclination. *Therefore. Proof of the Minor.* All motion that proceeds from man can be caused by God. But man can do something contrary to the natural inclination of a creature while the natural inclination of the creature remains, e.g., to throw a stone upwards, while the stone's natural weight remains. Therefore, God can bring it about that a stone is thrown upwards while the stone's natural weight remains. This is the case in other instances in which there is violent motion.

Confirmation. The motion whereby demons and souls are tormented is inflicted by God. But such a motion is especially violent, since it is especially afflictive. Therefore, God can introduce violence.

– *Objection*: The punishment inflicted on the demons is not properly speaking violent, but has only moral resistance. *Reply*: The punishment of the demons is sadness and affliction. But that which is saddening and afflictive is founded on some physical contrariety, namely in something which is contrary to nature itself. Therefore the punishment inflicted on demons is certainly physical violence. This is the common view of the Thomists.

SECOND QUESTION

How Nature is Ruled

We showed that there is a wondrous a dynamic and teleological order in the world.[25] But now we must discuss that order more specifically, that is, the laws that govern worldly things and the suspension of those laws that may be sometimes observed.

FIRST ARTICLE

ON THE LAWS OF NATURE

I. What is a Law of Nature. In the proper sense, law is defined as: "an ordering of reason to the common good promulgated by him who has care of the community" (*ordinatio rationis ad bonum commune ab eo qui curam communitatis habet promulgata*). Law, therefore, is essentially something that pertains to reason, even if positive laws presuppose an act of the will; it must be proposed for the sake of the community and ordered to the common good as to its end; its author is not a private person, but only he who has care of the community. Finally, promulgation belongs to the *ratio* of law, at least as a property [or as a *proprium,* a proper accident]. Law thus defined is taken formally as it is present in the ruler; it can be also considered materially and objectively, insofar as it is present in the directed and ruled thing.

Taken formally, law is not in nature, but in the Author or Governor of nature; but taken materially, it can also be attributed to nature, analogically and metaphorically. For natural causes seem to be directed *by some reason*, or by the ordering of reason, since they act according to a constant and uniform manner; and they also seem to tend *to the common good*, since they act for a universal end. On this account, the law of nature can be defined: "The *ratio* or principle whereby natural causes are steadily and uniformly directed" (*Ratio, seu principium quo causae naturales ad constanter et uniformiter agendum diriguntur*). – Therefore, the law of nature is not the very steadiness of effects that we observe, but the *intrinsic ratio* according to which this steadiness results.

[25] Treatise I, q. 2, a. 1.

II. How the law, order, and course of nature are distinguished. These three are often used interchangeably; but if used more precisely, they differ somewhat. We just explained what 'law' properly means. 'Order' is generally defined as, "the disposition of similar and dissimilar things that gives to each its place" (*parium dispariumque rerum sua cuique loca tribuens dispositio*). In the present context, it can be defined as, "the apt disposition of single things to attain their special end and of all things to attain their common end" (*apta disposition rerum naturalium ad assequendum finem sive specialem singulis, sive omnibus communem*). A twofold order of nature can be distinguished: (1) *the particular order of nature*, which depends on some cause, and contains those things that are subject to the cause—and this order is manifold according to the diversity of causes that act in creatures; (2) *the universal order of nature*, which depends on the first cause of all things, and embraces all things. The *particular* order is subject to the *universal* order; but the universal order is subject to the *most universal* order, which is the order of divine Providence.

'Course' signifies a swift and continuous movement from one place to another. For this reason the course of nature can be described as, "the succession according to which nature passes from one operation to another and then again" (*successio secundum quam natura transit de una operatione in aliam et ita porro*), without any interruption or cessation. It is commonly defined as, "the ordinary sequence or succession of natural effects, and their mutual dependence, according to natural laws." The course of nature, therefore, presupposes natural laws and occurs according to them.

Now, the sensualists, who believe that induction only gathers probable conclusions, and generally all philosophers who, like Hume, cast doubt upon the principle of causality or who deny final causes, reject the idea that there are physical laws in the world. Their principles are refuted in Logic or in Metaphysics.

On the other hand, evolutionists and determinists claim that the laws of the world are immune from all contingency and possess an insuperable necessity, such that they admit no exception. But the true teaching is developed through the following conclusions. Hence:

III. First Conclusion: "There are physical laws in the world."

1st Argument. A physical law, or a law of nature, is a principle of acting constantly and uniformly. But in the world there is a principle of acting constantly and uniformly. Therefore, a law of nature exists. *Proof of the Minor.* Wherever there is a cause that is *per se* determined to constant and uniform effects, there is a principle of acting constantly and uniformly. But in nature there are causes that are *per se* determined to constant and uniform effects; hence it is that we can predict the aforesaid effects in the future, and in fact do so through the certitude of science. Therefore, in nature there are principles of acting constantly and uniformly.

2nd Argument. All admit that there is a true and infallible science regarding many physical things. But the infallibility and certitude of the physical sciences is founded on the laws of nature. Therefore, the existence of the laws of nature cannot be cast into doubt. *Proof of the Minor.* The infallibility of science is founded on its object, or its immutable principle, which acts in a constant and uniform manner. But this principle of acting constantly and uniformly is what we call the law of nature. Therefore, the infallibility of the physical sciences is founded on the laws of nature.

IV. Second Conclusion: "Physical laws are contingent, absolutely and *simpliciter*; yet they are necessary hypothetically and *secundum quid*."

Proof of the 1st Part. The laws and order of nature cannot be understood without things, or substances, in which they are present. But all the things that exist in the world are *simpliciter* contingent.[26] Therefore, the laws and order of nature are subject to the same contingency.

Proof of the 2nd Part. The second part is established from those arguments through which we just proved the existence of natural laws. A law of nature is a principle of acting constantly and uniformly, and consequently implies a cause that is *per se* determined to uniform and constant effects. But a cause that is *per se* determined is a necessary cause. Therefore, a law of nature implies a certain necessity. Similarly, natural laws are the object of scientific certitude. Therefore, physical laws must be at least in some way necessary. This necessity,

[26] We amply proved this in Treatise I, q. 1, a. 2.

nonetheless, must not be confused with the necessity of metaphysical laws. Hence, it is established that:

V. – THIRD CONCLUSION: "The necessity of metaphysical laws is that [kind of necessity] which does not undergo any mutation or exception; whereas the necessity of physical laws does not exclude all exception or suspension."

Proof of the 1st Part. Metaphysical laws regard essences as far as their intrinsic constituents. But the intrinsic constituents of essences are altogether immutable, nor do they admit of any exceptions. Therefore, the laws of metaphysics are never suspended. Thus, no reason or cause can make it be the case that a man is not a rational animal, or that a triangle does not have three angles, or that two plus two does not equal four.

Proof of the 2nd Part. Physical laws regard either the natural operations of things, or their properties, or their essences themselves, not metaphysically and as far as their intrinsic constituents, but considered in their physical *esse*. But for none of these reasons is change or suspension impossible. Therefore, the necessity of physical laws is not such as to exclude all change or suspension. *Proof of the Minor, in parts*. (1) The natural operation of secondary causes depends on certain conditions, and especially on divine concurrence (*concursus*). But it is not impossible for the aforesaid conditions to be absent, or for God to refuse His *concurrence*: for just as God is not obliged to create or conserve things, so neither is He obliged to move or *occasion* them to act. Now, if the divine concurrence ceases, it is natural for the creature not to act, just as if the influx of a principal cause ceases, it is natural for the instrument to remain at rest. Therefore, it is not impossible for a natural operation to cease, e.g., for fire not to burn, if God removes his *concurrence*. Moreover, God can supply the efficiency and operation of creatures. For what nature can do, the Author of nature affords of Himself alone. God, therefore, can produce the effects of nature without the operation of nature, e.g., health without using the power of nature. Therefore, changes in and exceptions to [the laws of nature] are not impossible on the part of the operation of nature. (2) Even though properties flow naturally from an essence, nonetheless they are not so connected with it that they cannot be separated. They are like the effects of the essence; but if these effects are impeded, the essence remains the same and retains the same essential order to the effect: thus, if the actual extension of a body is removed, the body conserves its order to the actual extension; or if the weight of water is removed or impeded, the water retains the same relationship to the weight. Therefore, exceptions to and suspensions of [the laws of nature] are not impossible on the part of the properties.

(3) The *metaphysical* essences of things, as far as their essential predicates, remain altogether immutable; but *physically* and concretely considered, they are subject to some changes. First, they are mutable in that they can pass from *esse* into nothingness, as it is clear that a created essence can be annihilated by God. Second, one essence can be changed into another through a new form, as it is the case with mater-form composites. Spiritual essences, however, are certainly incapable of being generated or corrupted (*ingenerabiles et incorruptibiles*); but absolutely considered it is not impossible that, having removed all differences, God changes the being of one of them (*id quod est entitatis in una*) into the being of the other (*in id quod est entitatis in altera*), as will be proven, when we discuss obediential potency.[27] Therefore, change in and suspension of [the laws of nature] is not impossible on the part of essences, physically considered.

VI. – WHAT IS THE DIVISION OF PHYSICAL LAWS. The chief division is that into *particular* laws and *universal* laws (*leges universalissimae*). *Particular* laws are proper to some *ens* or to some determinate category of beings; whereas *universal laws* regard the whole of nature and the common good of the entire universe.

Further, universal laws are intended first and are the *ratio* for willing (*ratio volendi*) the particular laws. Universal laws, as we said, certainly regard the good and the common end of the whole universe. But the common end is the *ratio* for willing the particular ends; and the whole is the *ratio* for willing the parts, for in the order of intention, the whole is prior to the parts. Therefore, universal laws are the *ratio* for willing the particular laws. Hence, in the free decree whereby God established universal laws is included implicitly the decree concerning the particular laws which are included in universal laws.

Laws are further divided into cosmic, astronomical, physiological, chemical, etc., according to their proper subject matter.

---

[27] We show this in *Metaphysics: Ontology*, Treatise I, q. 2, a. 1.

SECOND ARTICLE

ON THE NATURE OF MIRACLES[28]

I. – Two Conditions for a Miracle. We just discussed the suspension of physical laws, or miracles. In the present article we inquire into the nature of a miracle; in the next, the possibility and knowability of miracles.

The term 'miracle' (*miraculum*), says the Angelic Doctor,[29] is taken from 'marveling' (*mirando*). Two things are required for marveling: first, that the cause be hidden; second, that the effect about which we marvel exhibits the opposite effect of what commonly occurs. Each time these two things concur, something marvelous or wondrous happens: thus, iron going upwards is an effect that is the opposite to that which commonly occurs; and, further, to many, the cause is hidden (*occulta*). Hence, such an effect, to some, is something *marvelous*. It is plain, however, that the aforesaid effect is not something *simpliciter* unusual (*insolitum*), or something that requires a cause that is *simpliciter* hidden, since it is known to many. Therefore, there are two kinds of wonder: some are *secundum quid* unusual and have a cause that is *secundum quid* hidden; but others are *simpliciter* unusual and have a cause that is *simpliciter* hidden. Those of the first kind are called marvels (*mira*) and illusions, whereas the latter may not only be called 'marvel in act' or 'marvel in potency', but also *miracles* (*miracula*), as having in themselves the cause of marveling.

Now, the event must be *simpliciter* unusual, not only beyond some particular order, but beyond the universal order of nature; similarly, a cause that is *simpliciter* hidden is that which is invisible, not only to the senses, but to all created or creatable intellect. But such a cause is God, whose essence is according to itself unknown to all created

---

[28] On miracles one may consult: St. Thomas, *Summa theologiae* Ia, q. 105; *Summa contra gentiles* 3.101; *De potentia* q. 6; the theologians in general in the treatise *De vera religione*; Hooke, *Theol. Curs.*, in Migne, t. 2, p. 507ff; Newman, *Essays on Miracles*; Monsabré, *Introduction au dogme catholique*, t. 2; Gondal, *Le surnaturel*, t. 2; De Bonniot, *Le miracle et ses contrefaçons*; *Le miracle et les science médicales*; Boissarie, who wrote on the miracles of Lourdes.

[29] Cf. *De potentia* 6.1.

or creatable intellect. Consequently, it belongs to the *ratio* of a miracle that it can only be caused by God.

II. – ST. AUGUSTINE'S AND ST. THOMAS' DEFINITIONS OF 'MIRACLE'. Augustine defines 'miracle' as, "something difficult and unusual that is above the power of nature and that appears beyond the expectation of him who marvels."[30] – It is said to be *difficult*, not with respect to God, who does it; nor on account of the dignity of the thing in which it happens, but relative to the power of nature. – *Unusual*, not because it rarely occurs, but because it is against the normal course of nature, even if it is repeated daily: thus, the transubstantiation of the bread into the Body of Christ occurs daily, yet does not cease to be a miracle. – *Above the power of nature*, because nature cannot in any way produce that effect, considered in its substance, or it cannot produce it in that subject, or at least it cannot produce it in such a way. *Beyond the expectation of him who marvels*, because the effect is opposed to that which should occur. Here it is understood that it is beyond the expectation of nature, although not beyond the expectation of grace.

In addition to this, St. Thomas gives his definition: "Those things are properly to be called miracles which are done by God beyond the order that is commonly observed in things."[31]

Which are done by God: the first condition, namely, that the cause be *simpliciter* hidden; *beyond the order...*, the second condition, namely, that the effect be *simpliciter* unusual.

We must now consider each condition.

III. – THE FIRST CONDITION IS CONSIDERED: "Beyond the order commonly present in things." It is not sufficient for the *ratio* of a miracle that it can be done by God alone; it is necessary that it be natural for them to act according to some customary way and, nonetheless, God makes them act beyond this usual way. Hence, creation and the *justification* of the impious are not miracles, because these, according their usual order, do not come about by any cause other than God. Further, it is not sufficient for the effect to be beyond the particular order of some nature, because then throwing a stone upward

---

[30] *De Trinitate* 3.5: "[A]liquid arduum et insolitum supra facultatem naturae et praeter spem admirantis apparens."

[31] *Summa contra gentiles* 3.101: "[P]roprie miracula dicenda sunt quae divinitus fiunt praeter ordinem communiter observatum in rebus."

would be a miracle, but must be beyond the order of every created nature. Hence, monstrosities are not miracles, because, although they are against a particular nature, they nonetheless do not occur against universal nature. The following condition, therefore, must be verified in the *ratio* of a miracle, that it be "beyond the order that is commonly present in things."

Now, two things are to be considered in this order: both the *things that are subject to the order*, and the *ratio* of the order, which is the divine preordination insofar as it includes a directive knowledge, a commanding will, and goodness as an end toward which all things are impelled. It is asked now whether a miracle can be said to be absolutely *beyond order*.

IV. – CONCLUSION: "A miracle cannot be said to be universally and absolutely beyond order: it is beyond order with respect to the things which are subject to order, but not beyond order with respect to the *ratio* of order."

That a miracle is beyond the order of things is clear from the above; that it is not beyond the *ratio* of order is easily proven. The *ratio* of order is a disposition, or preordination, of the knowledge and will of God, which directs all things towards His goodness as to their end. Hence, a miracle would be said to be against the *ratio* of order either because it is not willed by God, or because it is not comprehended by the divine knowledge, or because it is not ordered to the divine goodness.

But it is impossible for God to do something that He does not will, since creatures do not proceed from Him naturally, but through His will. Nor is it possible either that something be made by Him which his knowledge does not comprehend, since His will cannot will something unknown to Him. Nor is it again possible for Him to do something in creatures which is not ordered to His goodness as to its end, since His goodness is the proper object of His will.[32] Therefore, it is impossible that there be a miracle that is beyond the *ratio* of order.

Hence, a threefold order may be distinguished: the *particular* order of some nature; the *universal* order of all things; the *most universal* order of divine Providence. Now, a miracle is beyond any particular order and beyond the universal order; but in no way is it beyond the most universal order.

---

[32] Cf. *Ibid.* 3.98.

V. – Whether a Miracle is Something Against Nature. Many authors, like Bergier, define 'miracle' as, "a work against nature." Further, we must be aware that something can be said to be against nature in two ways: *secundum quid* and *simpliciter*. Something is against nature *secundum quid* when something occurs that is contrary to an inclination that exists in a thing; something is against nature *simpliciter* when that which occurs applies violence to nature. In the first way, we do not deny that certain miracles can be said to be against nature; and in this sense, the Angelic Doctor distinguishes between miracles *above nature*, miracles *against nature*, and miracles *beyond nature*. In the other sense, in no way can it be admitted that there can be a miracle against nature. Hence:

VI. – Conclusion: "A miracle *simpliciter* is not against nature." We shall prove the arguments that St. Thomas develops.[33]

1st Argument. That to which nature is in potency according to the natural order is not against nature, even if sometimes there is something contrary to a particular form, which is corrupted. Thus, when fire is generated and air is corrupted under the agency of the fire, this is a natural generation and corruption. But all creatures are in potency to being moved by God beyond the order of their own causes: for since God is pure act and all others bear some admixture of potency, it follows that God is to all things as something active is to that which is in potency. Therefore, a miracle, or that which God does in creatures beyond the order of their own causes, is not against nature, but natural.

2nd Argument. Creatures are instruments for whatever God may will. But it is not against the nature of an instrument that it be moved by the principal agent; rather, it is especially fitting, for it is for this very reason that instruments are made, that they may serve the action of the principal agent, when they are moved. Therefore, it is not against nature that a thing be moved in whatever way by God: for they are thus made so that they may serve Him.

3rd Argument. That is not against nature which is done through the action of an agent on which natural action depends: thus, the flowing back and forth of the sea, inasmuch as it is caused by the agent on which its natural inclination depends, is not said to be against nature, although it be beyond the natural motion of water. But a miracle is caused by the

[33] Cf. *Ibid.* 3.100.

action of the Agent on Whom the natural action and inclination of a thing depends. Therefore, a miracle is not against nature.

4th Argument. Since it is through the measuring of something that judgment concerning that thing is obtained, it follows that that through which something is conformed to the measure of a thing is said to be natural to that thing. But God, as the first *ens* which is to all things the cause of their *esse*, is the first measure of the essence and nature of anything. Therefore, that with which a thing has been endowed by God is natural to that thing. Therefore, even if God imprints something on a thing otherwise [than usual], it is not against nature.

5th Argument. All creatures are compared to God as artifacts to the artificer. But it is not against the *ratio* of an artificer if the artificer uses his artifact in a different way, even after giving that thing its first form. Nor is it, therefore, against nature if God uses something otherwise than according to the usual course of nature.

– We can also use the argument whereby we showed that God does not introduce violence in his creatures.[34] That is not against nature which is according to the more forceful inclination of a creature. But a miracle, and whatever God can do in a creature, is according to the more forceful tendency of the creature, namely, the inclination to obey the most universal Agent. *Therefore.*

VII. – The Other Condition for a Miracle is Considered, Namely, That It Can Only be Brought About by God. First of all, we concede that a miracle, broadly understood, due to the fact that it exceeds human faculties and consideration, can be done by the angels and demons; hence these marvels are not mere illusions, but sometimes true things, and true deeds. The angels can indeed move bodies locally and apply their activity on passive things; either way, effects that exceed human cognition and faculties result. The devil can, for example, move serpents with the utmost quickness from one place to another, and hence he can bring it about that serpents appear where no one suspected their presence; or to move a heavy body through the air so quickly that the body seems to fly, or to produce sound through the striking of the air. Similarly, by applying their activity to passive things, for example, by applying seeds and the fertilization of seeds by promoting their natural powers, the angels can produce serpents or frogs. Therefore, since angels know all the powers and hidden things of nature, they can

[34] Preceding question, a. 3, n. 8.

produce true marvels that exceed human cognition and faculties. But these are not miracles, properly speaking. Hence:

VIII. – FIRST CONCLUSION: "God alone can, as principal cause, work miracles."[35]

1st Argument. A miracle is an effect that is beyond the universal order of nature. But no creature can act beyond that order. *Therefore. Proof of the Minor.* That which is totally constituted under a given order cannot act beyond that order or come out of that order. But every creature is constituted under the universal order of nature. Therefore, no creature can act beyond the universal order of nature or come out of that order.

2nd Argument. A miracle consists in the fact that something occurs in a thing which is not within that thing's potency. But an angel or any creature can only bring about in a thing that which is within that thing's potency. Therefore, neither an angel nor any creature can produce a miracle. *Proof of the Minor.* Every creature in its action requires a subject upon which to act: but it belongs to God alone to do something, no subject being presupposed. But whatever depends on a subject for acting can only do that to which the subject is in potency: for the entire action of an agent upon a subject consists in reducing the subject from potency to act. Therefore, a creature can bring about in a thing nothing except that which lies within the potency of the thing.

3rd Argument. That which acts in virtue of nature does not work a miracle, since a miracle is beyond or above nature. But angels and any created artificer can only act upon nature in virtue of nature, since art, as it is clear from what was said above,[36] cannot introduce its action in nature except in virtue of nature. Therefore, angels, or any created agent, cannot produce a miracle.

IX. – SECOND CONCLUSION: "Nonetheless, good angels and men can be the instrumental causes of a miracle."

We said: "good angels," for it is not fitting to God to use demons as instruments of miracles. "Since miraculous action is a certain divine testimony that is indicative of divine power and truth, if the demons, whose entire will is turned to evil, are given any authority to work

[35] Cf. *Summa contra gentiles* 3.102.

[36] Preceding question, a. 3.

miracles, God would be a witness to their falsehood."[37] Having brought this to mind, we now prove the thesis.

Angels act in three ways in the working of miracles. First, by asking through their prayers; and this mode can be common to both angels and men. Second, by disposing matter by its natural power to be subject to a miracle, as it is said that in the resurrection angels will gather the dust of the dead, which will be restored to life by the divine power. The third way is when they work as cooperating, and this is instrumental causality properly speaking. Further, in order for rational creatures to be *physical* instruments of a miracle, it is required (and sufficient) that they receive from God some transient and fleeting power whereby they are subordinated to the principal cause. But nothing prevents rational creatures from being able to be endowed by God with this transient power. Therefore, creatures can be *physical* instruments of miracles. – The major is the notion of an instrumental cause. The agent must raise the instrument to his own level (*evehere ad se instrumentum*) through a certain power that is superadded, which is not a perfect form, but is something fleeting and transient, in the manner in which colors are in the air. The minor is also clearly established. It is not difficult for God to impress this power onto a spiritual creature, although he even uses corporeal creatures instrumentally for the sake of the justification of spirits, as is clear in the case of the sacraments. We should make an exception of those miracles that require creation, for a physical instrument of creation is a contradiction, as we have shown elsewhere.[38]

X. – Division of Miracles. Since a miracle is something that exceeds the power of nature, there are as many species of miracles as there are ways in which something can exceed the power of nature. Now, something exceeds the power of nature in three ways: *as far as its substance*, *as far as its subject*, and *as far as its mode*. Therefore, the division of miracles is threefold: *as far as their substance*, *as far as their subject*, and *as far as their mode*.[39] *Explanation of the Minor.*

---

[37] *De potentia* 6.5: "[C]um operatio miraculosa sit quoddam divinum testimonium indicativum divinae virtutis et veritatis,- si Daemonibus, quod quorum est tota voluntas ad malum, aliqua potestas daretur faciendi miracula, Deus falsitatis eorum testis existeret."

[38] Treatise I, q. 2, a. 4. – Cf. our work, *La causalité instrumentale en Théologie*, ch. 8.

[39] Cf. *Summa theologiae* Ia, q. 105, a. 8.

Something exceeds the power of nature *as far as the substance of the event*, when that which occurs can in no way be brought about by nature, as in the case of *transubstantiation*, in *the mutual penetration of bodies*, and in the *glorification of bodies*, and these are miracles in the highest degree. Now, something exceeds the power of nature as far as *the subject*, when that which occurs can be done by nature, but not in a given subject, like the *resurrection of the dead*, and *giving sight to the blind*: for nature can cause life, but not in a corpse, and can give sight, but not to the blind. And this is the second degree of miracles. And something exceeds the power of nature as far as *the mode and order of production* (*quoad modum et ordinem faciendi*), when that which occurs can be done by nature and in a given subject, but not in a given way, as when someone is *suddenly cured of a fever or of leprosy*, or when *suddenly the air is condensed into rain without natural causes*. And these hold the last place among miracles.

Elsewhere, the Angelic Doctor offers a different division,[40] which is nonetheless reducible to the previous division. Miracles are said to be *above nature*, *against nature*, or *beyond nature*. We must be aware that all miracles can be said to be *beyond* nature, since they are done beyond the order of all of nature; and they can be said to be *above* nature, since they exceed all the powers of nature; and we have already shown that no miracle is *simpliciter* against nature. The aforesaid division, therefore, is to be understood in a special sense. A miracle is said to be *above nature* when that which happens cannot be done by nature, either because the form that is introduced by God cannot be introduced by nature, as the form of glory; or, if nature can introduce that form, it cannot introduce it in that state, as life in the corpse. A miracle is said to be *against nature* insofar as there remains in the nature a disposition that is contrary to the effects that God operates, as when the Sun stands still, or when the sea was divided, or when fire does not burn, or when a virgin gives birth. Something is said to be *beyond nature* when nature can elicit the effect, but not in the way in which God produces it; thus the conversion of water into wine can be in some way done by nature (insofar as the water that is assumed by the vine as nourishment, in due time is turned into the juice of the grape through digestion), but it does not occur in an instant, as when Christ turned the water into wine.

Those miracles that occur *beyond nature* happen to be miracles for three reasons. First, *on account of their excess and their singular mode*

---

[40] Cf. *In II Sent*. d. 18, q. 1, a. 3; *De potentia* q. 6, a. 2, ad 3.

of production, as when a multitude of frogs were produced in Egypt. Second, because *they occur in a determinate time and hour*, as when someone is instantly cured upon the invocation of some saint. Third, *when something occurs universally*; as when, in the Old Law, it is said that after a drink of the water of jealousy,[41] by a divine miracle, the belly of the adulteress started rotting.

----

[41] Cf. Numbers 5:14.

THIRD ARTICLE

ON THE POSSIBILITY AND KNOWABILITY OF MIRACLES

I. – Opponents of Miracles. All pantheists, who confuse God with nature and think that the laws of nature are variable in infinite ways, believe that miracles are intrinsically contradictory. To these are joined both the deists and the rationalists, on the one hand, who either altogether attack the notion of divine Providence or have a perverted understanding of it; and, on the other hand, the materialists and the positivists, who claim that the laws of the universe are inescapable and are the necessary matter of evolution. Others, like Voltaire, admit that a miracle is not *simpliciter* impossible, but at least that it is less worthy of God. Others, finally, concede verbally the possibility of miracles, but they claim that miracles cannot be known with certainty or distinguished from natural effects. In this sense Renan said: "We do not say: 'Miracles are impossible'; we say: 'As of yet, no miracle has ever been established'."[42]

II. – The Possibility of Miracles is Examined On Three Accounts. Three elements are to be considered in any miracle: (1) the obediential potency of the creature, or the capacity that the creature has to be able to be raised to all the effects which the first Agent wills to produce in it; (2) the laws of nature, whose suspension is brought about through the miracle; (3) the Author of the miracle, namely, God. *Simpliciter*, therefore, a miracle is to be considered possible if there is no impossibility on any of these respects.

III. Miracles are not Contradictory on the Part of Obediential Potency. Since a creature is totally subject to the first Agent, its obediential potency extends to all things of which God is absolutely capable, namely, to all those things which by their definition and *essence* do not involve any contradiction. But the essence of a thing does not involve a contradiction in a miracle. *Therefore. Proof of the Minor.* The essence of a miracle does not consist in a substance becoming an accident, or an accident becoming a substance, or that a

[42] *Vie de Jésus*, Introduction: "Nous ne disons pas: Le miracle est impossible; nous disons: Il n'y a pas eu jusqu'ici de miracle constaté."

triangle consist in two angles, etc.; rather, it consists in the nature of some event, which is a *substance*, a *quality*, or a *change*: thus, in the resurrection of the dead, the essence is the restored union of soul and body; in the glorification of the body, the essence is a quality that renders the body incorruptible. Yet none of these things imply contradictions in their definitions. And such is the case with the rest also.

IV. – MIRACLES ARE NOT CONTRADICTORY ON THE PART OF THE LAWS OF NATURE. The laws of nature are *simpliciter* contingent, as was shown in article one. Therefore, absolutely speaking, they can be suspended and changed. Yet, a miracle does not properly eliminate or change physical laws, but only suspends a particular effect of some law; while the children are kept unharmed in the fire, the common law that fire burns is conserved, but its effect is prevented in that given case; while accidents remain without a subject, the universal law that accidents actually inhere persists, and *de facto* that law has force with respect to the rest of the accidents. Hence, a miracle is only the suspension of a particular effect, or the particular exception in the laws of nature. But it was shown above that not all exceptions or suspensions are excluded from physical laws. There would perhaps be a contradiction if miracles did violence to nature, and would be against nature. Further, it is established from the foregoing that, *simpliciter*, a miracle neither is against nature nor introduces violence in nature, but rather, is according to a more forceful and stronger inclination of nature.[43]

IV. – MIRACLES ARE NOT CONTRADICTORY ON THE PART OF GOD. God possesses the attributes of power, goodness, immutability, and wisdom. And on account of all of these, miracles are fitting to God in the highest degree.

(1) *On the Part of His Power.* A miracle occurs when God, without the aid of nature, elicits an effect that should be done by nature, or elicits in nature that which in no way can be done by nature. But the divine power can of itself introduce whatever nature elicits. In fact, an agent through his will can produce instantly and without an intermediary whatever effect does not exceed the agent's power; but the most perfect Artificer can produce whatever effect an imperfect artificer produces.

[43] Above, a. 1, nn. 5ff.

Now, God is the most perfect artificer, acting through his will, and not through the necessity of nature. Therefore, the lesser effects that are produced by inferior causes God can produce immediately without secondary causes. – Further, the divine power is capable of producing many things of which nature is altogether incapable. For the higher a power is, the higher the effect that it can produce. But the divine power is the power of the most universal cause, which infinitely exceeds created powers. Therefore, it can produce innumerable effects that nature is altogether incapable of producing.

(2) *On the Part of His Goodness.* Although the order with which things are endowed represents in its own way the divine goodness, nonetheless it does not reproduce it perfectly and adequately. Now, that in an exemplar which is not perfectly exhibited in one way, can again be participated in, in another way beyond that. Therefore the divine goodness can be reproduced and represented in some other way beyond this order. But an effect that is beyond the order of things is a miracle. Therefore, the divine goodness can, without injury, shine forth through a miracle.

(3) *On the Part of His Immutability.* God, by causing a miracle, does not change His decrees, but fulfills in time that which from eternity He decrees that He will do. The Angelic Doctor says: "God from eternity foresaw and willed Himself to do what He does in time. Therefore, He thus constituted the course of nature, that He nonetheless would preordain in His eternal will what He would at some time do beyond this course [of nature]."[44]

(4) *On the Part of His Wisdom.* In working miracles, God does not act as an imperfect agent without forethought who corrects the defect of His works; but, because the order with which things are endowed does not match up to the divine wisdom, the divine wisdom is not limited to this order, and therefore can operate beyond it.

VI. – OBJECTIONS. The difficulties that are usually brought up have been solved, both through that which we noted regarding the nature of miracles, and from the preceding arguments. Yet here we briefly explain many other difficulties.

---

[44] *De potentia* q. 6, a. 1 ad 6: "Deus ab aeterno praevidit et voluit se facturum quod in tempore facit. Sic ergo instituit naturae cursum, ut tamen praeordinaretur in aeterna sua voluntate quod praeter cursum istum quandoque facturus erat."

1st Objection: A miracle is beyond order. But the divine Wisdom cannot operate beyond order. Therefore, it cannot produce a miracle. ***Reply***: *I distinguish the major*: that a miracle is beyond order as far as the things that are subjected to order, I concede; but that a miracle is beyond order as far as the things that are subjected to the order of reason, I deny. *I contradistinguish the minor*: that God cannot act beyond order as far as the order of reason, I concede; but that He cannot act beyond order as far as the things that are subjected to order, I deny; for the order of things, taken in this way, does not match up to the divine wisdom, nor represents the divine goodness perfectly. The reply is clear from the preceding article.

2nd Objection: God cannot act against Himself. But by acting against nature, He would be acting against Himself. Therefore, He cannot act against nature. ***Reply***: The entire argument can be granted, for we have shown that a miracle is not *simpliciter* against nature. Therefore, even if it is conceded that God cannot act against nature, it cannot be concluded that God cannot produce miracles. Further, just as God is able not to make a creature, He is also absolutely able to act against nature, and to negate it and to destroy it.

3rd Objection: God cannot be the cause of evil. But evil is said to be that which is beyond order. Therefore, God cannot cause something beyond order. ***Reply***: *I distinguish the major*. That God cannot be the cause of moral evil, I concede; but that God cannot be the cause of physical evil, I deny; for physical evils that are reduced to order contribute to the *ratio* of the beautiful. *I distinguish the minor*. That moral evil is said to be that which is beyond the universal order, I concede; but that it is beyond some particular order, I deny. *I distinguish the conclusion*: that God cannot do something beyond the universal order, I concede; that God cannot do something beyond a particular order, I deny.

4th Objection: It is inconvenient that a greater good be dismissed for a lesser good. But in a miracle, a greater good is dismissed for a lesser good, namely, the good of the universe for a particular good. Therefore, it is inconvenient that a miracle occurs. ***Reply***: *I deny the minor.* In a miracle, the whole order of the universe, in which its good consists, is not taken away; rather, only the order of some particular cause to its effect is removed. Therefore, in a miracle, a particular good is dismissed for a greater good and a higher end.

VII. – The Knowability of Miracles. It remains for us to refute those rationalists who claim that miracles cannot be recognized or discerned with certainty.

Now, three things are required and sufficient for the knowability of a miracle: (1) That the existence and truth of that fact be established; (2) that it be known that the occurrence is not a natural event; and (3) that it be clear that the occurrence is not merely marvelous or a diabolical illusion. But these three can be established with certainty. Therefore, miracles are knowable and discernible.

***Proof of the Minor in Parts*:**

Part I. Even though a miracle requires a supernatural cause, *in itself* it is a physical, sensible event that is fit for all intelligences. But sensible events can be proved with certainty, either through one's own experience, if the event is present, or through human testimony, if the event was in the past. Now, human testimony, when it possesses all its due qualities, is altogether certain; and, although in itself it belongs to the moral order, it is ultimately reducible to metaphysical certainty: *There is no effect without a cause.*[45] Therefore, the existence and truth of an event that is presented as a miracle can be established with certainty.

Objection. The adversaries reply, following Hume, that human testimony concerning miracles is to be rejected altogether. For it is easier for a thousand witnesses to err than for an event to occur that contradicts the laws of nature: for it is only morally impossible for a thousand witnesses to err, but it is physically impossible for an event to exist which is contrary to the laws of nature. *Therefore.* ***Reply***: The adversaries commit a sophism in their hypothesis. They suppose that physical laws are absolutely necessary and that a miracle is absolutely impossible. We have already refuted this. It is certainly easier for a thousand witnesses to err than for metaphysical laws, which undergo no exception, to change; it is also easier for a thousand witnesses to err than for physical laws to be changed by a merely natural cause; but it is almost infinitely easier for the laws of nature to be changed by the First Cause than for a thousand witnesses to err. For indeed, it is neither physically nor morally impossible for God to change physical laws; rather, it is natural for God, who freely created these laws, to be able freely to suspend them. And yet, for a thousand witnesses to err is impossible both physically and morally and, indeed, to a certain degree

[45] *Major Logic*, Treatise II, q. 3, a. 7.

even metaphysically, because then there would be an effect without a cause. If a thousand witnesses agree on a certain point, it is metaphysically certain that there is a cause of that agreement, and such a cause can be assigned to nothing other than the evidence of the event. For this reason Hume's assertion is to be denied *simpliciter*.

The rationalists insist: While two or ten witnesses profess that a dead man has risen, thousands on the contrary assert that the dead do not rise. But the testimony of a few cannot prevail over the universal testimony of the human race. *Therefore.* ***Reply***: Here again a sophism is committed, because the testimony is not regarding the same object. For, what do the thousands of witnesses, or universally all men, assert? That the dead to not *naturally* or *generally* rise. But concerning the particular case, of which they are not witnesses, they are altogether silent. And what do the two or ten witnesses profess? That *some* dead man was *in a particular way* and *supernaturally* risen from the dead. That twofold testimony is perfectly harmonious, and each is based on evidence: for it is certainly evident that the thousands of witnesses are not in error when they assert that the dead *in general* do not rise, and it can also be evident that the two or ten witnesses, greater than any exception, cannot deceive or be deceived when they affirm that *one man* rose from the dead, for which they serve as an irrefutable proof.

Part II. In order to discern miracles from merely natural events, it is not necessary *positively* to know fully all the powers of nature and their efficiency; it is sufficient to know them *negatively*. And this is manifest to most people. Therefore, miracles can be discerned from natural events by most people. ***Proof of the Minor***. Such negative knowledge is obtained either from *that which* occurs, as the mutual penetration of bodies, or from the subject *in which* it occurs, as resurrection, or from *the manner in which* it occurs, as the healing of leprosy produced by a mere command.[46] In a general way, a negative knowledge of the powers of nature can be partly compared to metaphysical reasoning: e.g., we know that the effect that creation requires comes only from God; similarly, from metaphysics we know with certainty that God alone can be the Lord of life and death, and that only He can foreknow future contingents or free acts. It can also be compared to physical reasoning, on account of which it is clear that a corpse that already smells cannot naturally rise from death, otherwise it would be wrong to ask that it be buried. Similarly, from *that which* is

---

[46] Recall the discussion on the nature of miracles in a. 2.

done or from the *manner in which it is done*, the phenomena of imagination are distinguished from supernatural events.

Certain authors, with Renan, claim that miracles cannot be known with certainty except if they are produced before the scientific academy. This proposition (1) does injury to God's dignity, for God does not do great miracles in order to satisfy scientists, but only out of His freedom in order to make manifest His goodness. It also (2) goes against the nature and end of a miracle, because it is an extraordinary event, not to be judged by scientific laws, or to be confined in a laboratory, but must be made public and made suitable to all intelligences. (3) Scientists, on account of prejudices, pride, and passion, are not capable of judging divine events. For this reason, P. Monsabré very beautifully refutes Renan, a most vain man, showing "that he does not understand at all the dignity of God, the nature and end of miracles, the passions and habits of scholars, the weaknesses of our poor spirits, or anything at all of what he speaks."[47]

Part III. These are the signs whereby diabolical works can be discerned from true miracles: (1) *the nature and duration of works*; (2) *the life and customs of the worker*; (3) *the circumstances of the event*; (4) *the end* of the work (*finis operis*), whether proximate or ultimate.

Concerning the first, we must consider whether the work exceeds the powers of all of nature by reason of itself, or by reason of the subject, or by reason of its mode. It must be noted that the devil cannot elicit anything other than that to which the natural subject is in potency, for he does not have *creative* power, but only *eductive* power. Hence, if it were established that an event were greater than the natural potency of the subject, then it would be clear that it does not proceed from the devil. Further, the works of God are lasting and persistent, whereas diabolical illusions are generally temporary and gradually vanish.

Concerning the second, we know that, although God can, absolutely speaking, use a perverse man for the utility of others, nonetheless He generally chooses only good and holy men to produce miracles. One certainly cannot directly prove the falsehood of a miracle from the perverseness of the worker, but a great suspicion is aroused. But if, on the contrary, there is sanctity of life, at once it becomes

[47] *Introduction au dogme catholique*, 22e confér: "[Q]u'il n'entend rien à la dignité de Dieu, rien à la nature du miracle, rien à son but, rien aux passions et aux moeurs des savants, rien aux faiblesses de notre pauvre esprit, rien enfin à ce qu'il dit."

apparent that there is no fraud, and it is completely reasonable to believe that the miracle is true, for the evil one, namely, the devil, uses evil people. Hence, those who cooperate with diabolical illusions are the arrogant, and exhibit themselves as a spectacle.

Concerning the third, it must be observed whether the circumstances demonstrate the importance, honesty, and religiousness [of the event]. For diabolical marvels are done together with spells and magic words, which are often mixed with childish, ridiculous, absurd, dishonorable, and impious things.

Concerning the fourth, it must be recalled that the proximate end of a miracle is the utility, whether corporeal or spiritual, of men; whereas diabolical illusions either merely move man's curiosity or harm men, or, if they seem to have some utility, they are later used to another perverse end, namely, the spiritual detriment of the soul.

The ultimate end of the miracle is to promote virtue and sanctity, and finally the glory of God; whereas the end of diabolical illusions is to do injury to the faith, the denial of revealed truths (e.g., the spirituality of the soul, the eternity of punishments); and finally the eternal damnation of souls.

We omit here the other questions and difficulties that concern miracles, which are generally and more conveniently considered in the [theological] treatise *On the True Religion.*

Here, certain philosophers discuss mesmerism, magnetism, hypnotism, and ascribe these phenomena for the most part to the demons. But this consideration belongs more fully to the theologians and the moralists, since it relates to the intervention of angels or the liceity of doing these things; for this reason, we leave this topic to them, lest we diverge from the proper object of philosophy, although there a certain psychological aspect of hypnotism which we shall examine in Psychology.

THIRD QUESTION

What is Nature For?

We discuss now the finality of nature. Now, two things are asked regarding the finality of nature: (1) what is the end of nature; (2) whether this end consists in a certain indefinite progress and evolution, or, in other words, whether the end of nature is compatible or incompatible with evolutionism.

FIRST ARTICLE

THE END OF NATURE[48]

I. The Notion of 'End'. 'End' is defined as, "that for the sake of which something is done" (*id cujus gratia aliquid fit*), that is, that on account of which the agent acts. Now, an agent moves and is moved on account of some good; hence 'good' and 'end' are interchangeable (*bonum et finis convertuntur*), and each implies an order to the appetite. There is a distinction between the *finis qui* ('end which') and the *finis cui* ('end to which'). The *finis qui* is the good which the agent intends; the *finis cui* is the subject or person for whom that good is desired. The *finis qui* is twofold: *objective* and *formal*. The *objective finis qui* is the thing itself that is desired; the *formal finis qui*, also known as the *finis quo* ("end whereby"), is the possession of a thing, or the act whereby we attain the thing. We shall speak of all of these more fully in Ontology.

II. – Whether Nature Acts on Account of an End. The ancient materialists, Democritus, Empedocles, and Anaxagoras, denied that natural things tend toward a certain end. To these we must add the pantheists such as Spinoza, the evolutionists, the materialists, and the positivists. Against these is the following conclusion:

---

[48] Cf. St. Thomas, *Summa theologiae* Ia, q. 44, a. 4; *Summa contra gentiles* 3.17.

III. – CONCLUSION: "All natural things act on account of an end."

1st Argument. It is certain that natural things act. But no action is possible without an end. Therefore, natural things act on account of an end. *Proof of the Minor*: This is the law of subordinate causes: that if the first is removed, the others are necessarily removed. And the first of all causes is the end. Therefore, if the final cause is removed, all others are removed. Therefore, if the end is removed, all action becomes vain and impossible. ***Proof of the Minor of the Preceding Argument.*** The material cause and the formal cause depend on the agent [i.e., efficient] cause. And further, the agent [i.e., the efficient cause] depends on the final cause. Therefore, all causes derive their causality from the end, and therefore, the end is the first of all causes. ***Explanation of the Major of the Subsumed Argument.*** Matter does not obtain form by itself, just as nothing can reduce itself from potency to act: for this reason, matter must be reduced to form by an efficient cause. Therefore, matter and form are subject to an agent. But an agent, in turn, depends on an end. For unless an agent is determined to a certain effect, it would not bring about one work rather than another, since nothing follows from indifference. But to be determined to a certain effect is to depend on an end and to act on account of an end. Therefore, every agent depends on an end and acts on account of an end.

2nd Argument. If the works of nature fall short of the *ratio* of 'end', they happen by chance. But that all the works of nature happen by chance is altogether impossible, as has been proven more than once.[49] – Further, those things which proceed from chance do not happen constantly and uniformly. But in nature there are principles of constant and uniform action, which principles are the laws of nature. Therefore, nature does not operate by chance, but on account of an end. We shall again discuss these in Ontology.

IV. DIFFICULTY RESOLVED. To act on account of an end is to know the end. But natural agents cannot know their end. Therefore, they cannot act on account of an end. ***Reply.*** *I distinguish the major*: That to act on account of an end *formally*, directing oneself to the end, is to know the end, I concede; but that to act on account of an end materially and *executively*, as being directed by another, is to know the end, I deny. *I concede the minor. I distinguish the conclusion*: That

[49] Cf. Treatise I, q. 1, a. 2, n. 5; q. 2, a. 1, n. 1.

they do not act on account of the end formally, I concede; that they do not act on account of the end *executively*, I deny.

Something tends toward an end in two ways: *formally* and through itself, as man moves to some place through himself; *materially* and *executively*, when something is directed by something else toward an end, as an arrow is driven by archer to the target. Those things that tend toward an end in the first way know the end; but those who do so in the second way do not apprehend the end, although the end is known by the agent who directs them. For this reason, natural things are brought to their end as directed by the Author of nature, Who knows and appoints their end. Now, although they are directed by another, this impulse is not violent, as the impulse of the arrow toward the target; but those agents have in themselves some principle of impulse, by reason of which their inclination and impulse become natural to them. The arrow is driven by an extrinsic principle, whereas natural agents tend toward their end in virtue of an intrinsic principle, which is nonetheless directed by the Author of nature.

V. – The Ultimate *Finis Qui* of Nature is not Within the World, but rather is the Extrinsic Glory of God.

Proof. The order of agents corresponds to the order of ends; that is, a proximate and particular end is ascribed to a proximate and particular agent, whereas an ultimate and universal end is ascribed to a first and universal agent. But the first and universal Agent, namely God, is outside the world. Therefore, the ultimate end of things is outside the world, and this is God Himself. – And again, every being acts for the sake of some good, either to acquire it or to communicate it. But since God is the highest good and is most fully sufficient to Himself, He does not act for the acquisition of some good. Therefore, He acts for the communication of His own good. But the divine goodness, by the fact that it is communicated, it is manifested, known, and praised. Further, the manifestation and recognition of the divine goodness is the extrinsic glory of God, since glory is nothing other than 'splendorous recognition with praise' (*clara cum laude notitia*). Therefore, the extrinsic glory of God is the ultimate end of creation. One may consult the theologians regarding the intrinsic glory of God.

VI. – The Proximate *Finis Qui* is the Perfection of Creatures, and Especially the Beatitude of Rational Creatures.

For by the very fact that the divine goodness is communicated to creatures, creatures are assimilated to God. But the perfection of an effect consists in its assimilation to the cause. Therefore, the perfection of creatures is a result of the fact that they participate in the divine goodness; hence, this perfection is the proximate end of creation.

Further, irrational creatures, which are unable of themselves to know or to glorify God, are ordered immediately to the good of intelligent creatures. Hence, we can conclude that the good of rational creatures, that is, their happiness, is the principal proximate end of creation. God, therefore, disposed things in such a way that His glory consists in the happiness of creatures.

For the glory of God consists in the fact that He is known, praised, and loved by rational creatures; and the happiness of rational creatures consists in knowing, praising, and loving God.

VII. – On the *Finis Cui*. The ultimate *finis cui* is God, whereas the proximate *finis cui* is intelligent creatures. ***Brief Proof***. The *finis cui* is the person *to whom* the *finis qui* is attributed. But the person for the sake of whom the ultimate *finis qui*, the glory of God, is intended is God Himself; and the person for the sake of whom the proximate end, or the happiness of creatures, is obtained is rational creatures. Therefore, the ultimate *finis cui* is God, whereas the proximate *finis cui* is rational or intelligent creatures.

VIII. – The Proximate End of Nature Requires a Certain Connection Among Worldly Things.

Since worldly things coincide in their tendency toward the same end, they must be connected by some bond. Now, a bond can be understood in three ways: an *ontological* bond, or bond by reason of nature; a *dynamic* bond, or a bond by reason of causality; and a *teleological* bond, or a bond by reason of finality. We have sufficiently discussed dynamic and teleological bonds;[50] we discuss now the bond by reason of nature. This bond occurs in five degrees: (1) only *esse* in minerals; (2) *esse* and living in vegetables; (3) *esse*, living, and sensing in animals; (4) *esse*, living, sensing, and reasoning in man; and (5)

[50] Treatise I, q. 2, a. 1; a. 5, nn. 7ff.

perfect understanding in angels. All these things, according to the Scholastics, coincide in something common, yet they are essentially distinct. Some recent authors, after Bernoulli, Leibniz,[51] and Locke, posit a certain law of *continuity* that is present in all degrees of *ens*, such that in nature there is not even the least hiatus, but always between two species is found an intermediary species that participates in the properties of each of the other two.

IX. – The Law of Continuity can be Admitted in a Certain Sense, but Not in that Sense in which Recent Authors Construe It.

Explanation of the First Part. There is a Scholastic axiom: "The highest part of the inferior is attained by the lowest part of the superior" (*Supremum inferioris attingit infimum superioris*).[52] Organic and inorganic things consist of the same elements, and they are joined in such a way that the most perfect among inorganic things (e.g., crystals) imitate life, thus being assimilated to the lowest of organic things. They differ essentially from the *ratio* of life, however. Plants and animals coincide in some phenomena of life, and the more perfect among plants to some extent attain to the level of the lowest of animals that lack motion, yet they are essentially distinct by reason of sensation. The brute animal and man are joined by reason of sensitive life, and that which is highest in the sensitive order, namely the estimative power, to some extent attains to the level of that which is lowest in the rational order, namely, reasoning; for instinct, or the estimative power, imitates to some extent the judgment and reasoning of man, although there is great difference. By an outstanding capacity, the most perfect men are to some extent assimilated to the lowest angels, and the highest angels, namely, the seraphim, who are charity and love, are to some extent assimilated to the Holy Ghost, who is Love in God, although they are immensely distant from Him. Therefore, in a sense the law of continuity is to be admitted.

Proof of the Second Part. A species cannot possess properties that are mutually contradictory. But, if we admit the law of continuity in the sense in which recent authors understand it, the intermediary species would possess properties that are mutually contradictory, for these

---

[51] Cf. Leibniz, *Nouveaux essais* 3.6.

[52] Cf. Aristotle, *The History of Animals*, 8.1; St. Thomas, *Super Dionysium De divinis nominibus*, c. 8, lect. 30; our work, *Lumière et Foi*, c. 3.

intermediary species would participate in the properties of two opposite species. Therefore, these intermediary species are impossible.

Further, the proposition will be more fully evident from what we shall now say concerning the evolution of species.

WHETHER EVOLUTION IS IN AGREEMENT OR IN DISAGREEMENT WITH THE END OF NATURE

I. – What is Evolutionism? Since the proximate end of the world is the perfection of creatures, and especially rational creatures, we admit without hesitation a certain *progress* in virtue of which nature evolves more and more, and by which man further and further advances in the sciences and the arts. Yet, since every creature has a determinate mode [of being], the progression cannot occur infinitely unless species are destroyed and creatures themselves cease to be.

But many philosophers and scientists situate that perfection in a certain indefinite progress whereby nature is always raised higher, such that what was in the beginning in the state of [mere] matter later attains life through successive evolutions, arriving, from the lowest degree of life, to the ape, and from the ape to man, and in man himself it is continuously and constantly advancing. This hypothesis is called *evolutionism* or *transformism*. Further, universal evolutionism is fused together with the monism of the materialists, which claims that all beings, man included, arose out of primitive monera that were modified infinitely through mechanical powers. This is the view of Haeckel, Spencer, etc. We overthrew this monstrous error in the first treatise. Transformism is here restricted to the origin of species.[53]

II. – Opinions Concerning the Origin of Species.[54]

From the twenty-seven geological strata, it can be gathered that vegetative and animal organisms were successively changed through the destruction and renewal of species. Many species have gone extinct, such as the trilobites, the ammonites, reptiles of the secondary age, and many of the mammals that lived during the tertiary age and during the

---

[53] We shall speak later of the origin of life (Ia-IIae, Treatise I).

[54] Cf. Flourens, *Examen du livre de Darwin*; A. de Quatrefages, *C. Darwin et ses précurseurs*; de Nadaillac, *Le problme de la vie*; Jousset, *Evolution et transformisme*; Bianconi, *La théorie darwinienne*; Cochin, *L'évolution de la vie*; Gaudry, *Les ancêtres de nos animaux*; Blanchard, *La vie et les êtres animés*; Vigouroux, *Les livres Saints et la Critiq.*, tom. 2, append.; A. Farges, *La vie et l'évolution*; J. Guibert, *Origine des espèces*; de Kirwan, *Revue Thomiste* 9, pp. 379, 540; Elie Blanc and de Kirwan, *ibid.* p. 716; Abbé Leroy, *Pour et contre l'évolution.*

beginning of the quaternary age, such as the mastodons. How, therefore, does one species succeed another and, generally, what is the origin of species? The first solution is *creationism*, or *productionism*, which of old was the more common doctrine (*doctrina communior*) and was thought to be the more evident, and according to which God made each species, the perfect after the imperfect, out of inorganic matter. In this sense, Linnaeus says: "There are as many species as God created in the beginning." It would not be properly a creation out of nothing, but a unique drawing out of preexistent matter.

Another opinion claims that God used inferior species to generate superior species. This is the view especially of A. Gaudry. Some call this view *passive evolution by Divine influence.*

The third view is *active evolution*: God created in the beginning all species together, not in actual form, but virtually, or in seed form, as it were. Just as a mother is pregnant with her fetus, so the world itself is pregnant with the causes of new life.

Hence, primitive cells in this way virtually contained superior forms, such that, before they attained [the state of] perfect species, they had to go through intermediary forms and species. In this view of active evolution, divine influence can still be admitted, yet not immediately as in creationism or productionism. This opinion, which many modern Catholics hold, is attributed to Augustine; it is not refuted by St. Thomas, nor is it displeasing to Suárez.[55]

A fourth view is that of *passive evolution without divine influence*, but only with the aid of natural, irrational, and fortuitous causes.

Lamark claimed that the evolution of species occurred through adaptation to current conditions, or the *environmental conditions*, as they say: for, since individuals of the same species have lived in different circumstances, hence *variety* takes place, which is transmitted through heredity, and produces a new stock (or breed). Geoffroy Saint-Hilaire admits the fixity of species so long as external conditions remain intact, yet he claims that species necessarily evolves if conditions and circumstances vary. Darwin reduced all these theories into a system.

The principal points of transformism can be reduced to the following: *natural selection*, *struggle for existence*, *the law of heredity*, *adaptation to new circumstances*, *use and non-use*.

---

[55] Cf. ST. AUGUSTINE, *De Genesi ad litteram* 5; ST. THOMAS, *Summa theologiae* Ia, q. 66, a. 4; SUÁREZ, *De creatione*, disput. 15.

Darwin admitted, however, that God created some species. Having discussed these things, we now establish the following conclusion:

III. – CONCLUSION: "Purely Passive Evolution is to be Rejected Altogether."

Proof by going through each of the points of evolution.

(1) *Natural selection*. Man, they say, can perfect a species through artificial selection, by rejecting imperfect individuals, and conserving only perfect individuals. Nature similarly can produce more perfect species through a certain natural selection, by taking away imperfect animals and retaining the perfect. ***But we reply***: Regardless of what man can do with a species, he cannot produce a new kind. The result of [artificial selection] is variety; but, if the work and industry of man cease, gradually the variety ceases: for either the species becomes degenerate or it suffers from sterility. But, if a rational cause is not capable of obtaining excellent effects, what will nature, which is blind, obtain with blind and irrational causes?

(2) *The Struggle for Existence*. They say that animals are in a perpetual struggle for existence, and in this contest the weaker and less perfect perish. This is a postulate and a sophism. It is not a law intended by nature that individuals struggle against each other, but rather, it is a law of nature that all things aid each other, insofar as they are parts of the same universe that tend toward the same end. "We must help each other; that is the law of nature." Yet, if a conflict occurs *per accidens*, the species does not thereby change; rather, the result is only that the less perfect individuals are removed, and the species becomes purer and its purity is conserved.

Further, Darwin's claim that the struggle for existence *always* favors the strongest is not true; for in a struggle there are many fortuitous chance events that often in an astonishing way help the weak against the strong.

(3) *The Law of Heredity*. They propose a law whereby the varieties that are obtained through natural selection are fixed and stable. This law is also a postulate. It is clear, however, that through heredity only certain characters are transmitted. Deformities and accidental characters do not persist indefinitely; but nature returns with full force to the normal kind from which it had receded on account of a special accident or impediment. Thus nature would rather return to the primitive state. Therefore, this law of heredity would in fact preclude the absolute variation that the adversaries posit.

(4) ***Adaptation.*** We must pay special attention to external conditions, for they are not sufficient: they can aid the evolution of a preexistent power or aptitude, but do not make a new kind. In fact, quite diverse kinds are to be found in the same circumstances: thus in Europe there are various species under the same temperate conditions; twins in the same circumstances often are endowed with opposite qualities. Further, in different environments there are species that are altogether similar: the wolf, the fox, the fly, etc., exist in temperate, arid, and glacial regions.

(5) ***Use and Non-Use.*** We admit that an organ can be weakened, diminished, or to some extent cease to be, from non-use, and that an organ can evolve and be perfected from use; but from this it does not follow that a new organ comes to be or is created. If nature did this, why can man not grow new organs or new fingers for himself?[56]

IV. – THE ARGUMENTS THAT THE ADVERSARIES DRAW FROM GEOLOGY ARE NOT CONVINCING. For between different fossils there is often a gap, and one cannot find the intermediary beings that are necessarily presupposed in the theory of passive evolution. Contejean says: "One can challenge the transformists to cite a single example, any series of fossil types, where one can follow step by step, age by age, the metamorphosis leading from one species to another"[57] – And Blanchard says: "Today more than ever, I renew my call: it is with all the strength

---

[56] ABBÉ DE BROGLIE, *Religion et Science*, p. 202-203: "The spontaneous transformation of beings, without any superior cause; organs produced by needs that they excite and that they satisfy; blind animals acquire eyes because they desire to see; in another order, the spontaneous and causeless progress of intelligence and of the heart, which makes the genius of Plato and the soul of Saint Vincent de Paul arise out of the instincts and the sensations of a zoophyte.... What is this, Sirs, if not a new version of the old oriental cosmogonies, or even of Ovid's *Metamorphoses*?" (*Les transformations spontanées des êtres, sans cause supérieure quelconque, ces organes produits par les besoins mêmes qu'ils excitent et qu'ils satisfont, ces animaux aveugles qui acquièrent des yeux parce qu'ils désirent voir, et, dans un autre ordre, ce progès spontané et sans cause de l'intelligence et du coeur, quil fait sortir le génie de Platon et l'âme de Saint Vincent de Paul des instincts et des sensations d'un zoophyte, qu'est-ce, Messieurs, si ce n'est une nouvelle édition des vieilles cosmogonies de l'Orien, ou même presque des* Métamorphoses *d'Ovide?*).

[57] Cf. CONTEJEAN, *Revue scientifique*, 1881, t. I, p. 559.

of my soul that, at the beginning of this book, I submit this word to all the friends of the natural sciences: Show me once an example of the evolution of one species."[58]

If evolution and adaptation is a law of nature, it cannot happen that many animals fall short of it. Further, from geology it has been proven that many animals have fallen short of evolution and adaptation. *Therefore.* ***Proof of the Minor.*** Almost all species have remained unchanged and without variation from the quaternary age to our times, despite the variety of external conditions. Thus also, many organisms have not undergone the 'law' of evolution: species of the tertiary age, as the web-footed birds and many species of fish are still extant; among organisms of the secondary age, crustacean and zoophyte species; and even certain species of the primary age, e.g., some brachiopods, arachnids, etc.

V. – INDEED, GEOLOGY CONTRADICTS SUCCESSIVE EVOLUTION IN MANY RESPECTS. The different kinds of animals are often found, not successively, but together. Thus, in the primary age, the zoophytes, the segmented animals, and some vertebrates, such as the reptiles, existed together. Further, geology shows evidence that the more ancient animals were not always less perfect: for in the oldest strata are found fishes that possess a most perfect structure; in the secondary age, the reptiles are more perfect than the reptiles of the next age; many kinds of mammals of the tertiary age excel in many respects over the animals of our time. Therefore, geological facts in no way support successive evolution.[59]

---

[58] BLANCHARD, *La vie des êtres animés*, Préf.

[59] Hence, many scientists deny evolutionism, such as AGAZZIZ, DE QUATREFAGES, de NADAILLAC, BLANCHARD, etc. CONTREJEAN, *Revue Scientifique*, t. 7: "The proof of the metamorphosis from one species into another is still lacking: as we wait for this proof, the innumerable transformations invoked so far prove absolutely nothing in favor of the doctrine" (*La preuve de la métamorphose d'une espèce dans une autre est encore à désirer; en attendant qu'on la produise, les innombrables faits de transformation invoqués jusqu'ici ne prouvent absolument rien en faveur de la doctrine*). – H. FABRE, *Nouveaux souvenirs entomologiques*, p. 48: "I see big words; some cite natural selection, atavism, the struggle for life, but I would rather prefer the modest facts. After about forty years, I gather these modest facts and I interrogate them, and they do not quite reply in favor of the current theories" (*Je vois bien de grands mots, on invoque la sélection, l'avatisme, la lutte pour la vie, mais je préférerais quelques tout petits faits. Ces petits faits, depuis bientôt une quarantaine d'années, je les recueille, je*

VI. – Even if Evolution is Proven as a Fact, even then Divine Intervention would Not be Excluded. There is no effect that is greater than its principle. But the effect of orderly evolution is much greater than blind and irrational causes, for it belongs to the wise man to order. Therefore, if successive evolution could be proven as a fact, the intervention of an intelligent and wise cause, namely, God, the maker of species, would still have to be admitted. This is the common view of Catholic philosophers.

VII. – The Catholic Solution to the Problem of the Origin of Species. The three opinions at the beginning of this article, as recited by a Catholic, can be defended; of themselves they do not bear anything contrary to the faith. Each can be defended by means of arguments from suitability. They argue thus in favor of the second: It is fitting that God does not destroy species that already exist, but that he use them, by modifying them in order to draw out from them higher forms. In favor of the third: God only produces immediately those things that can only be made through His own action. But the production of species could have been caused by secondary causes, through the evolution of a primitive species in which the others were contained virtually. Therefore, it was more convenient that those species arise through active evolution.

Further, that opinion has the advantage that it coherently explains the facts that the evolutionists cite without any semblance of truth. For it can be defended by *embryology*, where it is observed that embryos of perfect animals go through certain phases in which they bear a similitude to inferior species, of which they are the last phase and terminus; it can also be defended from *paleontology*, which shows that the more perfect kinds existed after the imperfect; and it can also be argued from *anatomy*, which observes a similitude of kind among the diverse classes of animals.

Objection: It is metaphysically impossible that God communicate to a stone the power of reproducing, or to plants the power of sensing. Therefore, active evolution is metaphysically impossible. ***Reply***: It is certainly impossible for a stone, insofar as it remains a stone, to reproduce, and for the plant, insofar as it is a mere plant, to sense, just as

---

les interrogue, et ils ne répondent pas précisément en faveur des théories courantes).

the seed, insofar as it is a mere seed, does not generate a plant or animal. But it is in no way impossible for God to endow the stone with a certain power to evolve and to transition to the vegetative state, and finally to the form of an animal, as the seed, through an intrinsic power, evolves into a plant or a brute animal. Why could God not communicate what the seed has from a natural cause to the primitive monera?[60] Those philosophers go further, therefore, who posit an absolute and metaphysical impossibility of any evolution whatsoever.

Nevertheless, we defend the first opinion, which claims that God produced species immediately, as being *philosophically* more probable.

1st Argument. Although active evolution in the sense explained above does not involve any metaphysical impossibility, the fact of evolution should not be admitted unless it is established by *a posteriori* arguments. But *a posteriori* arguments do not prove evolution with certainty. For, indeed, the different kinds of animals are not always found to follow each other successively, but are sometimes found to exist simultaneously, and the imperfect do not always precede the more perfect. Now, even if it were proven that the perfect come *after* the imperfect, it could not be thereby concluded that the perfect come *from* the imperfect. Finally, many species from prehistoric times have remained intact and unchanged up until our times.

The similitude of kinds that is observed in embryology and in anatomy is general enough that it does not exclude many dissimilarities, both internal and external, among species, even species that are very closely related. Besides, it is only a sign that there is a unity of

---

[60] The analogy with the seed as a potential (and not) actual plant is scientifically inaccurate. For centuries, it was thought that seeds were not yet plants in act, but merely plants in potency. Today's biologists, however, have discovered that the seeds of plants are already (young) plants in act and not merely potential plants, insofar as they are already diploid organisms and not mere haploid organisms (i.e., they possess the complete genetic makeup of the parent plants, not only half of it, as is the case with the sex cells of the parent plants). Similarly, they regard human embryos as actual (young) human beings, and not as merely potential humans, insofar as they are diploid organisms with 46 (and not 23) chromosomes; the sperm and the ovum, however, are not considered by biologists to be actual human beings because they are haploid organisms (i.e., they only possess 23 chromosomes, which is half of the complete genetic code of the parent human being). Yet, despite the inaccuracy of the analogy, Hugon's argument is still sound: the analogy is meant as a mere illustration, and not as the crux of the argument. – ***The Translator.***

similitude among all species, as works of the same Author; it does not clearly prove that all species have come from the same primitive species.

2nd ***Argument***. If we were to admit evolution, we would have to conclude that the inferior species are ordered to the highest kind as to their proper and essential complement, and therefore all inferior kinds are something essentially transitory, imperfect, and incomplete. But this seems to contradict experience and reason. It contradicts experience because it is clear that many species in themselves are complete and independent and that they bear in themselves actions that *per se* reach their terminus, and that are not carried over to some higher appetite. It contradicts reason because, then, the beauty of the universe would be destroyed. For, indeed, the beauty of the universe consists in the *numerousness* of forms that are *per se* complete. But, if all the inferior species are essentially transitory and incomplete, then the *numerousness* of forms that are *per se* complete would be eliminated. Therefore, the beauty of the universe would be destroyed. The Angelic Doctor comments: "In the foresaid order, the order to which the *ratio* of divine Providence pertains, we say that the first [element] is the divine goodness, as its ultimate end, which is the first principle in acting, then the *numerousness* of things, for which are necessary the diverse degrees of forms and matters and agents and patients and actions and accidents. Therefore, just as the first *ratio* of divine Providence *simpliciter* is the divine goodness, *so the first ratio in creatures is their numerousness, and everything else seems to be ordered to the conservation and constitution of the fact that they are numerous*. And according to this, that which Boethius says at the beginning of his *Arithmetic* seems reasonable, namely, that all things that were constructed by the primaeval nature of things seem to have been formed by the *ratio* of number."[61]

---

[61] *Summa contra gentiles* 3.97: "In praedicto autem ordine, secundum quem ratio divinae providentiae attenditur, primum esse diximus divinam bonitatem, quasi ultimum finem, qui est primum principium in agendis; dehinc vero rerum numerositatem; ad quam constituendam necesse est gradus diversos in formis et materiis, et agentibus et patientibus, et actionibus et accidentibus esse. Sicut ergo prima ratio divinae providentiae simpliciter est divina bonitas, ita prima ratio in creaturis est earum numerositas, ad cuius institutionem et conservationem omnia alia ordinari videntur. Et secundum hoc rationabiliter videtur esse a Boetio dictum, in principio suae arithmeticae, quod omnia quaecumque a primaeva rerum natura constituta sunt, ex numerorum videntur ratione esse formata."

3rd Argument. Although the numerousness of species pertains to the nature of beauty, the production of species pertains to the constitution of nature itself. But the constitution of nature must proceed immediately from the Author of nature. Therefore, the constitution of species must come immediately from God. – ***Objection***: We ought not posit a miracle arbitrarily. But the immediate production of species by God would be a miracle. Therefore, the production of species is to not be ascribed immediately to God. ***Reply***: *I deny the minor*. That which God does in constituting nature are not miracles, since these things are not naturally done by anyone other than God. But the production of species pertains to the order, perfection, and constitution of nature. Therefore, the immediate intervention of God in the creation of species cannot be called a miracle, but rather, something natural and ordinary.

Here we conclude the first part of Natural Philosophy, giving great thanks to God, the Best and Highest, Whom we showed to be the Alpha and Omega, the Beginning and End of the world.